# Rechargeable Batteries
# Applications Handbook

# EDN Series for Design Engineers

# Rechargeable Batteries Applications Handbook

Technical Marketing Staff
of Gates Energy Products, Inc.

**Butterworth-Heinemann**
Boston   London   Oxford   Singapore   Sydney   Toronto   Wellington

Recognizing the importance of preserving what has been written, it is the policy
of Butterworth-Heinemann to have the books it publishes printed on acid-free
paper, and we exert our best efforts to that end.

Library of Congress Cataloging-in-Publication Data
Rechargeable batteries applications handbook / Gates Energy Products.
       p.      cm.
     ISBN 0-7506-9228-6 (case bound)
     1. Storage batteries—Handbooks, manuals, etc.    I. Gates Energy
Products.
TK2941.R43    1991                                           91-19522
621.31'2424—dc20                                                 CIP

British Library Cataloguing in Publication Data
Rechargeable batteries applications handbook.
   I. Gates Energy Products
   621.31242
   ISBN 0760692286

Butterworth-Heinemann
80 Montvale Avenue
Stoneham, MA 02180

10   9   8   7   6   5   4   3   2   1

Printed in the United States of America

# Contents

*Section* **3**  **Sealed Nickel-Cadmium Cells and Batteries   35**

# List of Figures

# List of Abbreviations

| Abbreviation | Description | Abbreviation | Description |
|---|---|---|---|
| A | ampere | KOH | potassium hydroxide |
| AC | alternating current | LED | light emitting diode |
| A/D | analog to digital | log | logarithm |
| Ah | ampere-hour | $m^2$ | square meter |
| °C | degree Celsius | $\mu A$ | microampere |
| **C** | charge or discharge rate as related to capacity | mA | milliampere |
| | | mAh | milliampere-hour |
| CC | constant-current | $\mu F$ | microfarad |
| cm | centimeter | mg | milligram |
| $cm^3$ | cubic centimeter | mm | millimeter |
| CP | constant-potential | MPV | mid-point voltage |
| $C_p$ | effective parallel capacitance. | mV | millivolt |
| | | mW | milliwatt |
| CPRV | cell polarity reversal voltage | Ni | nickel |
| | | Ni-Cd | nickel-cadmium |
| $\Delta T$ | (delta T) incremental temperature | NTC | negative temperature coefficient |
| $\Delta TCO$ | (delta TCO) incremental temperature cutoff | OCV | open circuit voltage |
| | | OEM | original equipment manufacturer |
| $\Delta V$ | (delta V) incremental voltage | | |
| | | P | pressure |
| dia. | diameter | Pb | lead |
| DC | direct current | psia | absolute pressure, pounds per square inch |
| dT/dt | rate of temperature change | | |
| DTC | dump timed-charge | psig | gauge pressure, pounds per square inch above atmospheric |
| dV/dt | rate of voltage change | | |
| $E_o$ | no load voltage | | |
| EOCV | end of charge voltage | PTC | positive temperature coefficient |
| EODV | end of discharge voltage | | |
| EMF | electromotive force | RAV | required application voltage |
| °F | degree Fahrenheit | | |
| F | farad | $R_e$ | internal resistance |
| g | gram | $R_T$ | thermal resistance |
| GEP | Gates Energy Products | s | second |
| h | hour | SLB | sealed-lead battey |
| Hz | hertz | TCO | temperature cutoff |
| I | current | V | volt |
| $I_{mp}$ | current at maximum power | VA | volt-ampere |
| IEC | International Electrochemical Commission | VCO | voltage cutoff |
| | | VDCO | voltage decrement cutoff |
| | | VLTCO | voltage limit temperature cutoff |
| IPCO | inflection point cutoff | | |
| in | inch | W | watt |
| $in^2$ | square inch | Wh | watt-hour |
| kg | kilogram | | |

*Section* $1$

# Introduction

As the world increasingly comes to rely on electrical and electronic systems for its daily functions, both vital and mundane, it also becomes increasingly dependent on batteries. As electrical and electronic systems become smaller and more efficient, batteries provide the key to portability. And, even with stationary systems, batteries are becoming increasingly important to provide memory backup or keep essential functions operational when the main power is not available.

Not only has the world become more dependent on batteries in general, it has also come to appreciate the economy and reliability of rechargeable batteries. Thanks to rechargeable batteries, travellers can now work at their laptop computer throughout a transcontinental airline flight and then recharge their batteries overnight using only a few cents worth of electricity. At home, rechargeable appliances kept on their chargers are always fully charged and ready for use. Rechargeable flashlights provide the security of knowing they will respond when needed — eliminating the mystery of when the last time the batteries were changed and guessing at how much they have been used since then.

Now, improvements in battery technology are opening many new uses for batteries. The rapid development of markets such as portable electronics and cordless appliances and power tools has relied heavily on today's improved batteries. In a diversity of applications, creative use of batteries is translating directly into improved sales and increased profits. Only recently, cold feet were the bane of many skiers; now, ski boots heated by rechargeable battery packs are keeping skiers happy and boot manufacturers scrambling to meet demand. But, creative use of batteries is aided greatly by knowing how batteries operate. This Handbook introduces the two leading forms of rechargeable batteries: sealed nickel-cadmium and sealed-lead batteries. Its objective is to help users and potential users of these battery systems understand how to best apply them. From this knowledge will come not only better designs for battery-powered equipment, but also creative new uses for batteries.

The Rechargeable Battery Application Handbook provides design and application information which is directly relevant to the sealed nickel-cadmium and sealed-lead battery families. This Handbook is designed to offer practical explanations of battery behavior with the focus on results rather than theory. The information included herein has been collected over the last quarter century by scientists, engineers, and technicians at Gates Energy Products.

## ORGANIZATION OF THE HANDBOOK

The Battery Application Handbook is designed for ready use by people at all levels of experience in batteries. For the newcomer, it offers an introduction to the basics of batteries, including their history, how they operate, and representative applications. The glossary at the back of the Handbook explains many terms relevant to batteries. For the experienced user, it provides in-depth discussions of charging, discharging, and other battery performance aspects important to designing a rechargeable battery into a particular application. The Appendices (Appendix-Sealed Nickel-Cadmium Cells and Appendix B - Sealed-Lead Cells and Batteries) contain a wealth of performance data and specifications for these battery types as produced by Gates Energy Products.

The one area that the Rechargeable Battery Application Handbook does not treat in great detail is choosing between sealed nickel-cadmium and sealed-lead batteries for a specific application or product design. Experience indicates that making this choice is either very easy or reasonably difficult. In the "easy" category are situations where only one type of battery is feasible. For example, only one chemistry is available at the needed size or the application is a retrofit designed to fit only one battery shape. If the choice is not self-evident for this sort of reason, then selection between the two forms of battery becomes more difficult. Both the sealed nickel-cadmium and sealed-lead batteries are excellent high-performance systems. They do, however, have a wealth of differences that have to be evaluated before choosing one over the other. Consultation with full-line battery manufacturers offering both technologies may prove useful in choosing the appropriate battery type for a specific application.

The remainder of the Handbook is organized as follows:

- Section 2 - Rechargeable Cells and Batteries
- Section 3 - Sealed Nickel-Cadmium Cells and Batteries
- Section 4 - Sealed-Lead Cells and Batteries
- Glossary
- Index
- Appendices

Section 2 is a general description of rechargeable batteries including a brief history of batteries, an introduction to some common battery terms, and discussion of the construction and theory of operation of nickel-cadmium and sealed-lead cells. It also provides a summary of the current status of development efforts for advanced rechargeable batteries that are beginning to enter the market.

Sections 3 and 4 cover application details for the sealed nickel-cadmium and sealed-lead battery systems respectively. The format within each section is identical. A brief introduction presents some of the features and benefits of the battery system. Then Sections 3.1 and 4.1 discuss discharge performance and how cell discharge performance may be affected by various use, design, and environmental factors. This is the information necessary to determine the basic suitability of the battery for the proposed application. After determining that the battery can meet the electrical load, it is time to develop the charging strategy best suited to the requirements of the application. Sections 3.2 and 4.2 cover various schemes for charging the batteries including discussions of charging efficiency, continuous and fast charging, and overcharging. Proper storage is covered next. Understanding how self-discharge may affect storage is important in designing a storage procedure that is compatible with the application. These concerns are covered in Sections 3.3 and 4.3. Sections 3.4 and 4.4 present methods of estimating battery life under various use patterns. Methods of prolonging life are also discussed here. Sections 3.5 and 4.5, the application sections, supplement the information presented in the earlier sections with application and design-related factors affecting battery operation. A brief discussion of battery use in specific applications is also included. The next sections (3.6 and 4.6) talk about quality control and test procedures: ways to ensure that the system design (battery and charger) is adequate to the duty and methods to ensure batteries are ready for use. The last sections (3.7 and 4.7) are devoted to safety and ensuring that batteries are used in a way that does not present a hazard.

Note that the information provided in the body of the Handbook (Sections 3 and 4) is fairly generic to the particular battery family being discussed. The curves shown in these sections are not usually dimensioned. Instead they are intended to provide general indications of cell and battery behavior in various situations. Detailed information on actual performance for specific cell types as manufactured by Gates Energy Products is provided in the appropriate Appendix. Appendix A provides performance data for sealed nickel-cadmium cells. Appendix B provides equivalent detail on the performance of sealed-lead cells and batteries.

## NOTICE TO READER

It is the responsibility of each user to ensure that each battery application system is adequately designed, safe and compatible with all conditions encountered during use, and in conformance with existing standards and requirements. The circuits contained herein are illustrative only and each user must ensure that each circuit is safe and otherwise completely appropriate for the desired application.

This Handbook and its appendices contain information concerning products manufactured by Gates Energy Products. This information is generally descriptive only and is not intended to make or imply any representation or warranty with respect to any product. Cell and battery designs are subject to modification without notice. All product descriptions and warranties are solely as contained in formal offers to sell or quotations made by Gates Energy Products.

*Section 2*

# Rechargeable Cells and Batteries

As electrical and electronic devices become increasingly essential parts of modern society, we are ever more dependent on our sources of electrical power. Batteries are one of the few practical methods of storing electrical energy. As such, they are vital components in electrical and electronic devices ranging from portable electrical shavers to satellites in space. Recent advances in battery technology, both in new battery types and in improvements to existing batteries, have fueled a surge in battery applications. As battery applications become more diverse and more critical to system operation, it is especially important that system designers and users understand the fundamentals of battery function.

This section provides a general introduction to rechargeable batteries before the Handbook splits into more specific discussion of the two major rechargeable battery types that are widely available today. It begins with a brief history of battery development and then talks about some general concepts important to batteries. The section then presents the chemistries and construction details of nickel-cadmium, nickel-hydrogen, and sealed-lead battery cells. It then briefly discusses some recent developments in rechargeable battery technology and usage.

# 2.1   Rechargeable Battery History

Batteries of one form or another have existed for nearly 200 years. From the beginning, researchers have been attempting to improve the energy density and make battery packaging more convenient for the user. This development work continues today because market opportunities expand immensely with each significant improvement in battery performance.

## 2.1.1   EARLY WORK

Most historians date the invention of batteries to about 1800 when experiments by Alessandro Volta resulted in the generation of electrical current from chemical reactions between dissimilar metals. The original voltaic pile used zinc and silver disks and a separator consisting of a porous nonconducting material saturated with sea water. When stacked as sketched in Figure 2-1, a voltage could be measured across each silver and zinc disk. Experiments with different combinations of metals and electrolytes continued over the next 60 years. Even though large and bulky, variations of the voltaic pile provided the only practical source of electricity in the early 19th century.

Johann Ritter first demonstrated a rechargeable battery couple in 1802, but rechargeable batteries remained a laboratory curiosity until the development, much later in the century, of practical generators to recharge them.

## 2.1.2   DEVELOPMENT OF LEAD BATTERIES

In 1859, Gaston Planté developed a spirally wound lead-acid battery system. His cell used two thin lead plates separated by rubber sheets. He rolled the combination up and immersed it in a dilute sulfuric acid solution. Initial capacity was extremely limited since the positive plate had little active material available for reaction. As repetitive cycling resulted in an increased conversion of the lead in the positive plate to lead dioxide, the capacity increased materially. This formation process remains a significant aspect of lead-acid battery manufacture today.

About 1881, Faure and others developed batteries using a paste of lead oxides for the positive-plate active materials. This allowed much quicker formation and better plate efficiency than the solid Planté plate. This improvement in battery technology occurred just as central-station electrical generation was becoming practical. One result of these two events was development of a diversity of commercial uses for lead-acid batteries including such applications as central stations, telephone exchanges, and train and residential lighting.

The next major influence on lead-acid battery development was Charles Kettering's invention in 1912 of the first practical self-starter for automobiles. General Motors subsequent adoption of battery-started cars provided the key for massive growth in use of lead-acid batteries. The use of lead-acid batteries in automotive starting, lighting, and ignition (SLI) service remains their largest market.

Although the rudiments of the flooded lead-acid battery were in place in the 1880's, there has been a continuing stream of improvements in the materials of construction and the manufacturing processes. Today, flooded lead-acid batteries exist in a variety of configurations tailored to the requirements of specific applications.

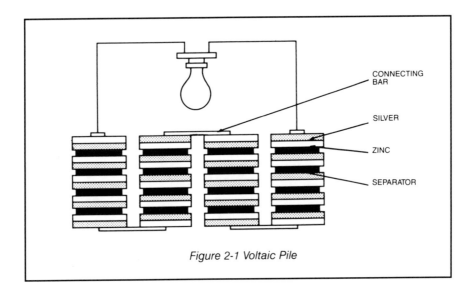

*Figure 2-1 Voltaic Pile*

Thanks to improved design and improved manufacturing quality control, present-day batteries have eliminated concerns over reliability and performance that prevailed as recently as twenty years ago.

Even recognizing the improvements in flooded batteries, they still contain liquid sulfuric acid with attendant safety, handling, and transportation concerns. Investigators have attempted a variety of approaches to immobilize the electrolyte in lead-acid batteries. The first effective result was the gelled electrolyte battery developed in Germany in the 1960's. This used silica gel in the electrolyte to greatly increase its viscosity. The result was a battery that substantially reduced concerns about leakage and spillage.

Working from a different approach, Gates Energy Products developed and was awarded a basic patent (U.S. Patent 3,862,861) on a sealed-lead battery. Using only minimal amounts of electrolyte and recombining evolved oxygen, the Gates sealed-lead battery provided major improvements in both performance and ease of use over both flooded and gelled-electrolyte batteries.

## 2.1.3   DEVELOPMENT OF NICKEL-CADMIUM BATTERIES

Development of practical rechargeable batteries using alkaline electrolytes lagged about 50 years behind lead-acid technology. About 1900, Edison began experimenting with a nickel storage battery with the goal of developing a practical electric automobile. In 1910 Edison demonstrated a commercial battery that used a nickel positive electrode, an iron negative electrode, and a potassium hydroxide electrolyte. Because of its ruggedness and high cycle life with repeated deep discharges, the Edison battery found commercial success in a variety of applications in the U.S. As nickel-cadmium batteries have become more cost competitive, nickel-iron batteries have lost most of their markets.

At about the same time as Edison's work in the U.S., Waldmar Jungner was working with first the nickel-iron and then the nickel-cadmium couple in Sweden. The result of his efforts was the pocket-plate nickel-cadmium battery which found widespread application in Europe, especially in larger sizes for stationary applications.

During World War II, the Germans developed the sintered-plate nickel-cadmium battery offering exceptionally high energy densities when compared with other rechargeable batteries. The sintered-plate, vented or flooded nickel-cadmium battery has found primary use in those applications such as aircraft engine starting where high performance will command a price premium.

In the 1950's, European experimenters developed a revolutionary form of nickel-cadmium battery that recombined gases evolved on overcharge instead of venting them. This closed cycle allowed them to develop a sealed cell with excellent performance characteristics. Because of its cleanliness and high energy density, the sealed nickel-cadmium cell continues to find broad application in electronics and consumer products.

## 2.1.4    RECENT DEVELOPMENTS

Interest in new and improved batteries remains strong today. The demand for versatile, clean, high-power energy sources grows as electronics becomes an increasingly essential part of both consumer and industrial products. To date, the results of the battery industry's development efforts have been most evident in the dramatic improvements in existing battery types. However, three new battery types, using different materials and technology, are beginning to find application.

Batteries using lithium metal offer the combination of high voltage and high energy density, although at a premium price. Lithium batteries are finding increasing application in situations that require the battery to supply a low drain rate for a long period. Although currently available as a nonrechargeable battery, rechargeable lithium batteries are beginning to investigate commercial markets.

A long cycle life combined with a high energy density has been the goal of the nickel-hydrogen development program which has focused on spacecraft applications. Here the complex design and attendant high cost are less important than high performance. Nickel-hydrogen cells are now viable competitors to nickel-cadmium cells in this very specialized market. See Section 2.4 for further discussion of the nickel-hydrogen system.

Finally, the system which currently is exhibiting the greatest application in commercial products is a hybrid of the nickel-hydrogen and nickel-cadmium technologies called the nickel-metal hydride system. Here absorption of hydrogen within a metal alloy's structure provides the energy source which powers the cell. Additional information on the nickel-metal hydride technology and other recent developments is provided in Section 2.6.

# 2.2   General Battery Concepts

As with any field, battery specialists have developed terminology and usage conventions that are routinely employed within their community, but that may be new or different to outsiders. This section describes some common terms and definitions that are often used in describing batteries or in defining battery applications.

## 2.2.1   PRIMARY VS. SECONDARY BATTERIES

Batteries are either primary or secondary. Primary batteries can be used only once because the chemical reactions that supply the current are irreversible. Secondary batteries, sometimes called storage batteries or accumulators, can be used, recharged, and reused. In these batteries, the chemical reactions that provide current from the battery are readily reversed when current is supplied to the battery.

Primary batteries are the most common batteries available today because they are cheap and simple to use. Carbon-zinc dry cells and alkaline cells dominate portable consumer battery applications where currents are low and usage is sporadic. Other primary batteries, such as those using mercury or lithium-based chemistries, may be used in applications when high energy densities, small sizes, or long shelf life are especially important. In general, primary batteries have dominated two areas: consumer products where the initial cost of the battery is very important and electronic products (such as watches, hearing aids, and pacemakers) where drains are low or recharging is not feasible.

The secondary battery, since it can be recharged, has traditionally been most heavily used in industrial and automotive applications. Here users were willing to trade higher initial cost and additional handling and care requirements for high current delivery and the economies of a rechargeable product. Only two rechargeable battery chemistries, lead-acid and nickel-cadmium, have, to date, achieved significant commercial success. The recently introduced nickel-metal hydride couple currently shows promise of supplementing nickel-cadmium cells in many commercial applications. Other couples, notably nickel-hydrogen and silver-zinc, are in use in special applications where performance requirements are more important than cost.

Recently, the battery market has changed significantly with consumers indicating that, for many products, they are willing to pay the initial premium necessary to obtain the long-term benefits of a rechargeable battery. Wet (flooded) versions of rechargeable batteries had achieved only limited penetration of the consumer and portable markets. Development of sealed high-performance forms of both nickel-cadmium and lead-acid batteries has allowed secondary batteries to make substantial inroads into traditional primary battery markets such as consumer products.

Recent improvements in secondary battery technology have improved performance and reduced costs. These improvements have spurred sustained growth in applications, both consumer and industrial, relying on rechargeable battery power. This trend is demonstrated by the penetration of rechargeable batteries into traditional primary battery markets such as flashlights. But the improved energy densities and cost/performance ratios of rechargeable batteries have also resulted in a wealth of battery applications (such as portable vacuums and power drills for consumers) that did not previously exist.

## 2.2.2   BATTERIES VS. CELLS

In casual usage, a battery is anything that supplies electrical power through chemical reactions. However, in discussing battery design it is important to understand the distinction between batteries and cells. Cells are the basic electrochemical building blocks. Batteries consist of one or more cells.

Cell voltage is determined by the electrochemistry involved. Nickel-cadmium cells nominally produce about 1.2 volts per cell while lead-acid batteries produce about 2 volts per cell. Battery voltages then must be multiples of the basic unit. For example, nickel-cadmium battery voltages may be 1.2, 2.4, or 3.6 volts, but not 3.0 or 4.5.

A battery can be a single cell provided with terminations and insulation and considered ready for use. More often, a battery is an assembly of several cells connected in series or parallel and with electrical output terminals (Figure 2-2). In many batteries, such as those used in automobiles, all intercell connections are made internally. The number of cells within the battery may not even be apparent from the finished package.

Often in discussions, the terms battery and cell are used interchangeably. The meaning is normally apparent from the context.

### 2.2.2.1   Cell and Battery Voltage

The voltage performance characteristics of a battery scale directly with the number of cells in the battery. This means that the voltage obtained on discharge or the voltage required on charge is usually just the appropriate single cell voltage multiplied by the number of cells in the battery. If a 2-volt sealed-lead cell requires a charge voltage of 2.4 volts, a 6-volt (3 cell) battery requires a charge voltage of 3 $\times$ 2.4 volts or 7.2 volts, a 12 volt (6 cell) battery will require 14.4 volts, etc. In discussions of battery output and charging, voltages are often presented on a per cell basis. These normalized values can then be converted to the appropriate voltages for a specific battery by multiplying by the number of cells in the battery.

### 2.2.2.2.   Cell and Battery Capacity

While the voltage of a cell is determined by its chemistry, cell capacity is infinitely variable. The capacity of a cell is essentially the number of electrons that can be obtained from it. Since current is the number of electrons per unit time, cell capacity is the integration of current supplied by the cell over time. Cell capacity is normally measured in *ampere-hours*.

The capacity of a cell is generally determined by the quantity of active materials (discussed below) included in it. Individual cells range in capacity from fractions of an ampere-hour to many thousands of ampere-hours. Cell capacity is usually directly related to cell volume, i.e. bigger cells usually mean higher capacities.

Especially for sealed nickel-cadmium products, certain cell sizes have become established as industry standards. This means that they conform to accepted dimensions for diameter and height. Much of the recent excitement and increased interest in rechargeable products has come from improvements in cell packaging and active materials utilization that provide higher capacities within accepted dimensional constraints.

Balancing the amounts of positive active material, negative active material, and electrolyte in a cell is one of the cell designer's tools. Adjusting the relative quantities of these three items allows the cell to be tailored for a specific application.

For many applications, a battery's ability to supply power (current $\times$ voltage) or energy (current $\times$ voltage integrated over time) may be of great concern. Instead

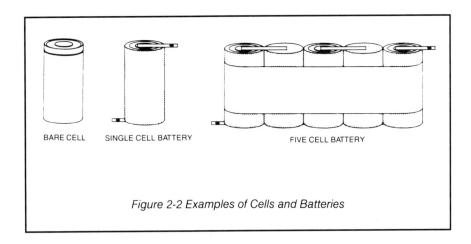

BARE CELL    SINGLE CELL BATTERY              FIVE CELL BATTERY

*Figure 2-2 Examples of Cells and Batteries*

of worrying strictly about how many amp-hours a battery needs to deliver, the designer may be concerned about how many watt-hours the cell will supply. Then, not just cell capacity, but also the ability to supply current at a specific voltage is an issue. This may be especially important in comparing batteries of different chemistries: a nickel-cadmium cell supplies current at about 1.2 volts while a sealed-lead cell operates at about 2.0 volts. Thus, for the same cell amp-hour capacity, the energy delivered may vary greatly. Current and power or energy delivery are discussed in greater detail in Sections 3 and 4.

### 2.2.2.3   Connecting Cells to Form a Battery: Series vs. Parallel

Every cell has a positive and a negative terminal. The terminals may be specific sites such as the positive and negative tabs on the sealed-lead cell or may be more general locations such as the positive cover and negative can on the sealed nickel-cadmium cell. In connecting multiple cells into batteries, there are two options: the positive from one cell may be linked to the negative of the succeeding cell (series connection) or the cell positive terminals may be linked together and the negative terminals may be linked together (parallel connection).

Series connection means that the voltages of the connected cells add while the capacity remains constant. So the battery voltage becomes the cell voltage multiplied by the number of cells and the battery capacity is the capacity of the individual cell. Thus to obtain a nominal 12 volt DC output, a battery might consist of 10 nickel-cadmium cells in series or 6 sealed-lead cells. Series strings are the most common means of connecting cells.

Batteries employing parallel connections are used when a higher capacity than that provided by the individual cell is needed. When possible, it is ordinarily less costly and more reliable to use a cell with a higher capacity than it is to connect multiple cells in parallel. However, situations do arise where larger cells are either unavailable or unsuitable, so a battery consisting of parallel strings of cells is used. In such cases, the battery voltage is that of the individual cell while the battery capacity is the individual cell capacity multiplied by the number of cells in the battery. Parallel connection of nickel-cadmium cells requires special charging considerations.

Occasionally, batteries rather than cells may be connected in parallel. The voltage and capacity results are analogous to the single cell case. Sections 3 and 4 discuss some special concerns regarding such series-parallel batteries.

Batteries should never be constructed from other than identical cells. Mixing of cell types, cell sizes, cells from different manufacturers, or, even, manufacturing lot may give results ranging from simply unsatisfactory to disastrous.

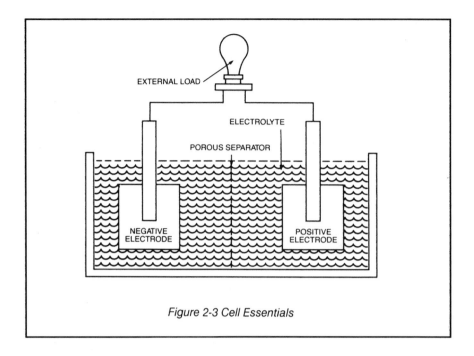

*Figure 2-3 Cell Essentials*

### 2.2.3    CELL COMPONENTS

The cell, the basic unit of the battery, has four main components as sketched in Figure 2-3. These are:

The *negative electrode* (the anode during discharge reactions) supplies electrons to the external circuit when oxidized during discharge. For the nickel-cadmium battery, the active material for the negative electrode is metallic cadmium, while metallic lead fills the same function for lead batteries.

The *positive electrode* (the cathode during discharge reactions) accepts the electrons from the external circuit when reduced during discharge. The active materials for the positive electrode are nickel oxyhydroxide ($NiOOH$) and lead dioxide ($PbO_2$) for the nickel-cadmium and sealed-lead batteries respectively. Because of their typical shapes, the electrodes are commonly referred to as the positive or negative plates. Both terms, electrode and plate, normally encompass the active material and any substrate used for support and for current collection.

The *electrolyte* completes the circuit internally by furnishing the ions for conductance between the positive and negative electrodes. The electrolyte can be either an alkaline solution which supplies negative ions ($OH^-$) or an acidic solution which provides positive ions ($H^+$) to conduct current. Charge flows from positive to negative electrode in two manners. In an alkaline electrolyte, negative ions are created at the positive electrode and absorbed at the negative. In an acidic electrolyte, positive ions are created at the negative electrode and absorbed at the positive. In either case, the effective flow of current is the same. The nickel-cadmium couple uses an alkaline electrolyte — a dilute mixture of potassium hydroxide ($KOH$) and water. The lead battery uses an acidic electrolyte — a dilute mixture of sulfuric acid ($H_2SO_4$) and water. In lead batteries, the sulfuric acid is consumed as the cell discharges. Thus, the acid concentration is a measure of the state of charge of the cell. This concentration (or specific gravity) variation is the operating principle behind the hydrometers used to indicate state of charge for flooded lead-acid batteries. The electrolyte in nickel-cadmium batteries acts only to convey ions; it does not participate in the reaction. As a result, monitoring state of charge through electrolyte concentration changes is not possible for nickel batteries.

A *separator* is normally used to electrically isolate the positive and negative electrodes. If the two electrodes make direct electrical contact, they form an internal short-circuit, discharging the battery and rendering it useless. Strictly speaking, the separator is not necessary since physical isolation can provide electrical isolation. However all commercial batteries use a separator to allow closer electrode spacing without creating internal shorts. The type of separator used varies by cell type. Sealed nickel-cadmium cells use a porous plastic separator while advanced versions of the sealed-lead cell use a porous glass-fiber separator. In these examples, both nickel-cadmium and sealed-lead, the separator also absorbs electrolyte, limiting the amount of free electrolyte found in the cell and keeping the electrolyte next to the electrode.

These components are then housed in a cell jar or can. Depending on the type of battery, a variety of other components may be used to package the cell, to support the various components, or to provide for easy electrical connection. But only the four components listed above are essential to cell function.

## 2.4  CLASSIFICATION OF APPLICATION TYPES: FLOAT OR CYCLIC

For many aspects of battery performance, a key parameter is the amount of time that a battery discharges compared with the time it spends on charge. Although battery applications may fall anywhere along the spectrum of charge time vs. discharge time, many applications cluster at the two extremes.

Given the reliability of the AC electrical power in most developed countries, batteries that provide backup power in case of AC failure usually spend the vast majority of their life on charge. These are float applications. Here the major concern is battery life when subjected to a continuous trickle charge.

By contrast, in cyclic applications the battery is discharged regularly. In this type of service, the principal concern is that the battery recharge quickly compared with the discharge time. Many portable applications are cyclic. For example a miner normally uses his battery-powered lamp for the duration of an eight-hour shift and then wants the battery charged overnight and ready when he begins his shift the next day.

The distinction between float and cyclic applications proves to be very useful in discussing both battery and charger selection. Obviously, there are many mixed uses that carry aspects of both float and cyclic duty. In evaluating a battery and charger combination for a particular application, the specific duty scenario should be carefully considered and, where possible, testing performed to that scenario. But, in more general consideration of batteries and chargers, the distinction between float and cyclic applications proves very useful.

## 2.2.5  THE METHOD OF SPECIFYING CHARGE AND DISCHARGE RATES: THE C RATE

Different sizes (capacities) of the same cell family often behave similarly in responding to charges or discharges that are scaled by the cell's nominal rated capacity. A discharge at the 1 amp rate means vastly different things to a 1 amp-hour capacity cell compared to a 10 amp-hour cell, but a 1 amp-hour cell will generally respond to a 0.1 amp discharge in the same general way that a 10 amp-hour cell responds to a 1 amp discharge. The scaling parameter for current flow rates is the C rate which is defined as the current flow rate that is numerically equal to the cell rated capacity. Thus for a 1 amp-hour capacity cell, the C rate is 1 amp while the C rate is 2.5 amps

for a 2.5 amp-hour cell. Care must be taken, however, to understand how the cel
actually rated. Nickel-cadmium cells are usually rated at the 1-hour or 5-hour ra
while sealed-lead cells and batteries are usually rated at the 10 or 20-hour rates.

In general scaling on capacity is very effective within families of cells. Thu
is used throughout this Handbook to present information on current flows. Both
charge and discharge rates are normally represented as multiples of the $C$ rate.

Some care should be used in applying $C$ rate scaling to high current dischar
At rates of $10C$ or more, cell and battery performance is highly sensitive to details
design and construction so even nominally similar cells may behave quite differen

# nium Cells
# es

s a unique set of desirable physical and electrochemical
ade it the system of choice for many applications. These
o basic types: sealed and vented. The sealed cell operates
rmally allowing no escape of gas, while the vented cell
n the cell as part of its normal operating behavior.
ed nickel-cadmium cell is a clean, rugged, lightweight
und diverse applications, especially in the consumer and
ial form of the nickel-cadmium cell with construction
ut hermetically sealed has been widely used in some
ications.

## M CHEMISTRY

ell may vary, the basic chemistry of the couple remains the
n cell is an electrochemical system in which the active ma-
ctrodes change in oxidation state without any deterioration
tive materials are present as solids that are highly insoluble
Unlike many other systems, the nickel-cadmium cell
does not require the transfer of material from one electrode
s are long-lived, since the active materials in them are not
on or storage.
um cell, nickel oxyhydroxide, NiOOH, is the active materi-
plate. During discharge it reduces to the lower valence
i(OH)$_2$, by accepting electrons from the external circuit:

$$+ 2H_2O + 2e^- \overset{Discharge}{\underset{Charge}{\rightleftarrows}} 2Ni(OH)_2 + 2OH^- \qquad (0.490 \text{ volts})$$

the active material in the charged negative plate. During dis-
nium hydroxide, Cd(OH)$_2$, and releases electrons to the

$$Cd + 2OH^- \overset{Discharge}{\underset{Charge}{\rightleftarrows}} Cd(OH)_2 + 2e^- \qquad (0.809 \text{ volts})$$

during charging of the cell.
ccurring in the potassium hydroxide (KOH) electrolyte is:

$$_2O + 2NiOOH \overset{Discharge}{\underset{Charge}{\rightleftarrows}} 2Ni(OH)_2 + Cd(OH)_2 \qquad (1.299 \text{ volts})$$

## L-CADMIUM CELLS

m cell operates as a closed system that recycles the gases cre-
s eliminates the electrolyte loss that occurs in vented cells. As
e several advantages over vented cells. The most obvious ben-
elimination of the need for electrolyte replenishment. Elimin-
enance requirement results in lower operating costs, improved
freedom in battery location. Additional benefits of sealed cells

include elimination of corrosion from vente¢
orientation.

Although most commercial lines of sea
vent mechanism as a safety measure, they are
out the safety vent, misapplication and abuse
sure that might damage the cell. The safety ve
safely releases gas (oxygen and/or hydrogen)
reseals, allowing the cell to continue operating
cells are hermetically sealed since charge con€
specialized applications.

## 2.3.2.1    Theory of Sealed-Cell Operation

Sealed nickel-cadmium cells normally o
the vent pressure because gas evolved during c
pliance with three essential design criteria is n€
criteria are:

1. The capacity of the negative electrode mu
   electrode, i.e. the negative electrode must
   than the positive. As a result, the positive
   emits oxygen before the negative electrod
   which cannot be readily recombined.
2. The electrolyte must be uniformly distribu
   across the surface of the two electrodes. T
   cell must be only enough to wet the plate s
   tor. The electrodes are not immersed in a p
   flooded cell.
3. Oxygen gas must be free to pass between t
   sufficient open area in and around the sepa
   able to diffuse efficiently from the positive

In charging a sealed cell the positive electr
negative electrode. At this stage additional charg
potential to rise until all the incoming current is €
ing oxygen gas at the positive electrode:

$$2OH^- \rightarrow \frac{1}{2}O_2$$

The oxygen generated at the positive electrode di
trode where it is reduced back to hydroxyl ions:

$$\frac{1}{2}O_2 + H_2O + 2e^- \rightarrow 2OH^-$$

The hydroxyl ions complete the circuit by migrati
Thus, in overcharge all the current generates oxyg
bined. The oxygen pressure initially increases but
pressure determined by cell design, the ambient te
this equilibrium state the generated oxygen is recc
the negative electrode never becomes fully charge¢

The efficient recombination of oxygen enab
ly overcharged at specified rates without developi
can thus be kept on trickle charge, maintaining the
for long periods.

It is important to remember that during charg
is an exothermic reaction. When the cell is in overc

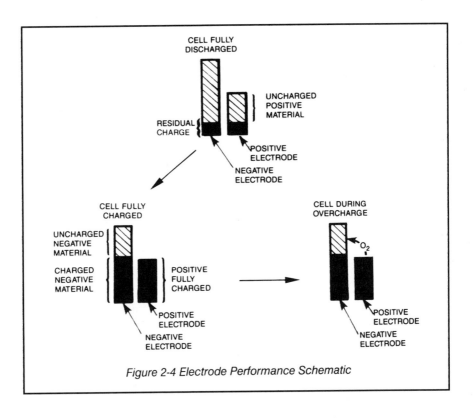

*Figure 2-4 Electrode Performance Schematic*

in the current coming in is converted to heat. With proper matching of the charger to the battery and attention to battery location and heat dissipation, the battery system can be designed to reach a thermal steady-state that will permit the battery to remain in overcharge indefinitely.

Figure 2-4 shows the electrode design of a typical sealed nickel-cadmium cell. When the cell is fully charged, the positive electrode active material has been converted to the high valence form, NiOOH, but the negative electrode still contains uncharged active material, $Cd(OH)_2$. In overcharge, oxygen gas is evolved from the positive electrode. Overcharge and its implications for battery design and charger selection is further discussed in Section 3.2.

## 2.3.2.2  Sealed-Cell Construction

A typical sealed nickel-cadmium cell uses a wound-plate, sealed construction with a nickel-plated steel can as the negative terminal and a metallic cell cover as the positive terminal. The cell cover is an assembly that includes the high pressure safety vent mechanism and insulating ring shown in Figure 2-5.

Each electrode, which is a continuous conductive strip containing active material, is isolated from the other electrode by the separator, a nonconducting, porous, fibrous, polymeric material. The two electrodes and their accompanying separators are wound together into a roll configuration shown in Figure 2-6. This roll is then inserted into the can so that the negative electrode contacts the can. A terminal from the positive electrode provides the electrical connection to the cover. An insulating seal ring isolates the positive cover from the negative can and provides a gas-tight seal for the container. Before sealing the can, a precisely metered amount of electrolyte, enough to wet the plates and provide an electrical path between them while still providing room for oxygen diffusion, is injected. Figure 2-7 shows a cutaway of a typical assembled cell.

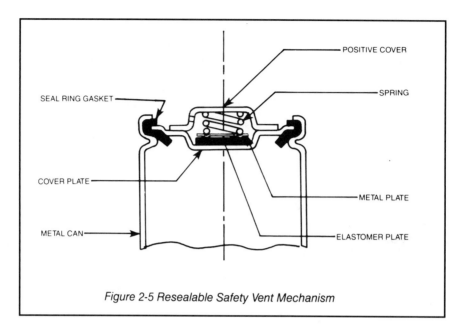

*Figure 2-5 Resealable Safety Vent Mechanism*

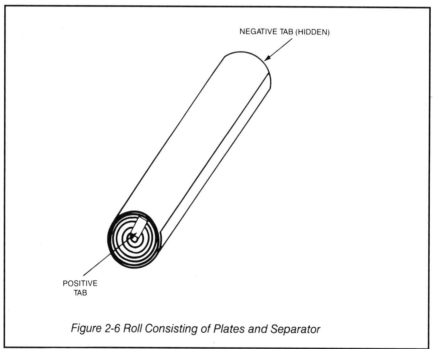

*Figure 2-6 Roll Consisting of Plates and Separator*

### 2.3.3    AEROSPACE CELLS

Sealed nickel-cadmium cells are also used for aerospace applications. Although these cells operate on the same principle as the commercial cell, their construction is necessarily somewhat different because of the environment in which they operate. As shown in Figure 2-8, these cells are rectangular, containing flat plate electrodes. Considerable research and development effort has gone into the design of the electrodes and the manufacturing processes for producing aerospace cells. They meet the stringent requirements necessary for operation in space vehicles. They are lightweight,

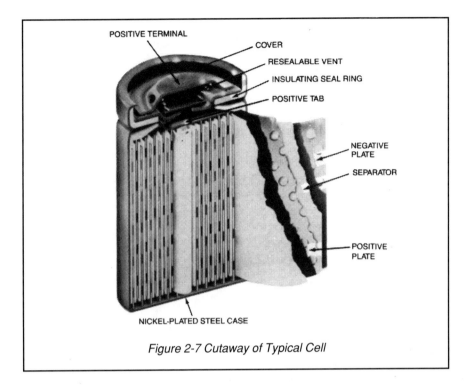

*Figure 2-7 Cutaway of Typical Cell*

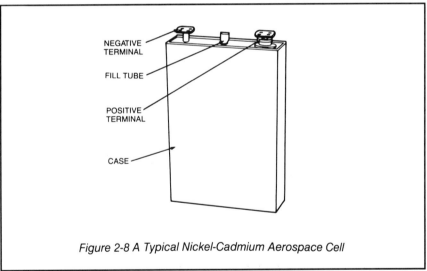

*Figure 2-8 A Typical Nickel-Cadmium Aerospace Cell*

capable of many charge/discharge cycles, and, above all, highly reliable and free from defects that may affect the operating life of the space vehicle.

## 2.3.4   OTHER FORMS OF NICKEL-CADMIUM BATTERIES

While sealed cells are today the dominant form of nickel-cadmium battery, there are other vented battery forms that find application in larger battery sizes. Vented nickel-cadmium batteries have some of the highest energy densities of any commercially available rechargeable batteries. They can deliver exceptionally high currents for their capacity. Their power qualifies them for demanding industrial applications such as starting aircraft turbine engines. Other applications for vented nickel-cadmium batteries include telecommunications and large uninterruptible power supplies.

Two types of vented nickel-cadmium batteries are commonly found today. Both are flooded designs that do not recombine the gases formed during overcharge. The difference between the two is the way the active materials are supported on the electrodes.

In sintered plate designs, the electrode support and charge collector consists of steel with a porous sintered nickel coating on it. The active materials for the positive and negative plates (nickel hydroxide and cadmium hydroxide respectively) are impregnated into this coating. This construction gives a very rugged battery with low electrical resistance that is particularly well suited for the high vibration environments on board aircraft.

Pocket-plate nickel-cadmium batteries take a very different approach to mating active materials to charge collectors within the electrodes. The charge collector consists of perforated steel sheet formed into pockets. The active materials are then fed into the pockets in bulk form, either as pressed briquettes or as a powder. This design eliminates many complex manufacturing steps giving lower costs per amp-hour than other nickel-cadmium batteries, but at a penalty in energy density. Thus pocket plate batteries are often found in large stationary applications such as uninterruptible power supplies. With proper attention in these uses, the pocket-plate nickel-cadmium battery is an exceptionally long-lived power source.

# 2.4   **Nickel-Hydrogen Cells**

Lightweight high-performance nickel-hydrogen cells are increasingly popular as the backup power source for space vehicles. The nickel-hydrogen cell is something of a hybrid between a battery and a fuel cell combining many of the features of a nickel-cadmium aerospace cell with aspects of a hydrogen-oxygen fuel cell.

The advantages of the nickel-hydrogen system are:

- high energy density
- a long cycle life, even in deep cycling applications
- high tolerance for electrical abuse (overcharging and reversing)
- state of charge indication through $H_2$ pressure

These advantages must be weighed against the cost and complexity of the cell. To be effective, these cells must maintain a hydrogen pressure approaching 1000 psi. Developing lightweight pressure vessels capable of containing hydrogen at these pressures requires innovative design and careful attention to details of materials selection and manufacturing techniques.

These cells use a nickel positive electrode that is similar to those used on aerospace nickel-cadmium cells. The electrolyte is potassium hydroxide just as with the nickel-cadmium cells. But the negative electrode consists of gaseous hydrogen contained at high pressure.

The pertinent reactions are:

At the positive electrode:

$$NiOOH + H_2O + e^- \underset{Charge}{\overset{Discharge}{\rightleftharpoons}} Ni(OH)_2 + OH^- \qquad \text{(0.490 volts)}$$

At the negative electrode:

$$\tfrac{1}{2}\, H_2 + OH^- \underset{Charge}{\overset{Discharge}{\rightleftharpoons}} H_2O + e^- \qquad \text{(0.828 volts)}$$

For the overall cell reaction:

$$\tfrac{1}{2}\, H_2 + NiOOH \underset{Charge}{\overset{Discharge}{\rightleftharpoons}} Ni(OH)_2 \qquad \text{(1.318 volts)}$$

During overcharge oxygen generated at the nickel electrode and excess hydrogen recombine to form water using a platinum electrode to catalyze the recombination. The efficiency of this recombination process is one of the reasons that the nickel-hydrogen cell is so tolerant of overcharge.

The first aerospace nickel-hydrogen battery was launched in 1977. Since then, nickel-hydrogen battery performance has been highly satisfactory. Nickel-hydrogen batteries have been powering the INTELSAT-V family of communications satellites since 1983.

Figure 2-9 shows the external configuration of a typical nickel-hydrogen cell.

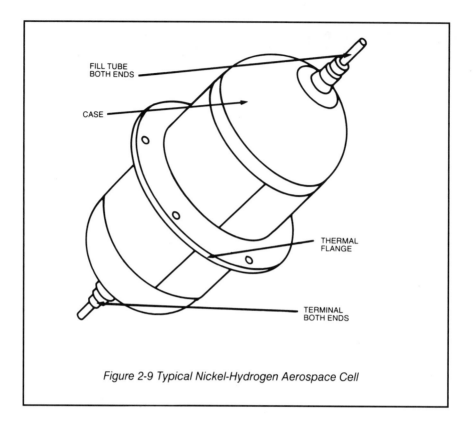

FILL TUBE
BOTH ENDS

CASE

THERMAL
FLANGE

TERMINAL
BOTH ENDS

*Figure 2-9 Typical Nickel-Hydrogen Aerospace Cell*

# 2.5  Sealed-Lead Cells and Batteries

The sealed recombining form of lead-acid batteries first appeared in commercial use in the early Seventies. It offered the designers of many systems needing secondary batteries a clean, economical, high-performance alternative to existing choices. Today, sealed-lead cells are operating effectively in many markets previously closed to lead batteries. Wound sealed-lead products provide many unique performance advantages that make them an excellent complement to sealed nickel-cadmium batteries.

## 2.5.1  ELECTROCHEMISTRY OF THE LEAD CELL

All lead-acid batteries operate on the same fundamental reactions. As the battery discharges, the active materials in the electrodes (lead dioxide for the positive electrode and sponge metallic lead for the negative) react with the sulfuric acid in the electrolyte to form lead sulfate and water. On recharge, the lead sulfate then converts back to the lead dioxide and metallic lead. Since the sulfuric acid is "consumed" in the discharge reaction, measurement of acid concentration in open or flooded cells, either through pH or specific gravity provides an indication of the discharge state of the cell.

The basic reactions involved are:

At the positive electrode:

$$PbO_2 + 3H^+ + HSO_4^- + 2e^- \underset{Charge}{\overset{Discharge}{\rightleftarrows}} PbSO_4 + 2H_2O \qquad \text{(1.685 volts)}$$

At the negative electrode:

$$Pb + HSO_4^- \underset{Charge}{\overset{Discharge}{\rightleftarrows}} PbSO_4 + H^+ + 2e^- \qquad \text{(0.356 volts)}$$

For the overall cell:

$$PbO_2 + Pb + 2H_2SO_4 \underset{Charge}{\overset{Discharge}{\rightleftarrows}} 2PbSO_4 + 2H_2O \qquad \text{(2.041 volts)}$$

## 2.5.2  SEALED-LEAD CELLS

Although the governing reactions of the sealed cell are the same as other forms of lead-acid batteries, the key difference is the recombination process that occurs in the sealed cell as it reaches full charge. In conventional flooded lead-acid systems, the excess energy from overcharge goes into electrolysis of water in the electrolyte with the resulting gases being vented. This occurs because the excess electrolyte prevents the gases from diffusing to the opposite plate and possibly recombining. Thus electrolyte is lost on overcharge with the resulting need for replenishment. The sealed-lead cell, like the sealed nickel-cadmium, uses recombination to reduce or eliminate this electrolyte loss. The design techniques used by Gates Energy Products to minimize electrolyte loss within the sealed-lead cell are covered by U.S. Patent 3,862,861. The theory of sealed-lead cell operation and its implementation in Gates cells are explained below.

### 2.5.2.1  Theory of Sealed-Cell Operation

As a discharged lead-acid cell starts to recharge, the vast majority of the electrical current input to the cell goes to convert lead sulfate to metallic lead at the negative and lead dioxide at the positive electrode. As the cell approaches full charge, the chemical conversion process becomes less efficient and increasing amounts of the current input are going to the electrolysis of water. Once the cell is fully charged, essentially all the current is going to electrolysis. At this time, hydrogen gas is being evolved from the negative electrode and oxygen gas is being evolved from the positive electrode. In a flooded lead-acid battery in overcharge, these gases bubble off with consequent water loss. The Gates sealed-lead cell uses techniques similar to those used in sealed nickel-cadmium cells to recycle the gases and prevent water loss.

Development of a recombining sealed-lead cell required attention to both the electrochemistry of the cell and to the design of the cell itself. The key to the process is minimizing the evolution of hydrogen gas at the negative electrode since the hydrogen, once evolved, does not readily recombine. Thus, Gates sealed cells are designed so that the positive electrode is limiting. This means that, as the cell is being charged, the positive electrode reaches full charge before the negative. As the charging process continues, the positive plate begins to evolve oxygen from the reaction:

$$2H_2O \rightarrow 4H^+ + O_2 + 4e^-$$

The oxygen then diffuses to the negative plate where it is readily recombined by the reaction:

$$Pb + HSO_4^- + H^+ + \tfrac{1}{2} O_2 \rightarrow PbSO_4 + H_2O$$

This has the effect of discharging the negative electrode slightly so that it can accommodate overcharge.

As long as the overcharge current remains low (0.01C or less), the charge and recombination reactions can remain in approximate equilibrium and little gas is generated. If recharge rates exceed this level as the battery approaches full charge, then substantially more oxygen evolves than can be recombined and the result is cell venting. Avoiding this condition is the goal of the various charging methods described in Section 4.2.

In order to maintain efficient recombination, three things must occur:

- the mass balance between the plates must ensure that the positive plate reaches full charge before the negative,
- the cell must allow ready diffusion of oxygen gas from the positive to the negative electrodes, and
- the charging rate must not be excessive.

The first consideration is the province of the electrochemist, the second is determined by the cell internal construction, and the third is the responsibility of the designer of the end-use system. Aspects of the sealed-lead cell construction, especially as they affect enhancing oxygen diffusion, are explained in section 2.5.2.2.

## 2.5.2.2 Construction of Sealed-Lead Cells and Batteries

The appearance of the wound sealed-lead cell is more like a sealed nickel-cadmium cell than it is like other forms of lead batteries. Unlike other lead-acid batteries that use multiple plates, Gates wound sealed-lead cells consist of only two (one positive and one negative) plates. Each plate consists of a lead grid coated with a paste containing lead compounds. When the battery receives its first charge, the materials in the paste are transformed into the active materials needed by the cell electrochemistry. Glass fiber separator electrically isolates the plates. The two plates and their separators are wound together in the spiral configuration shown in Figure 2-10. Once the two plates are wound together, the resulting element is then packaged in one of two forms: as a single cell such as shown in Figure 2-11 or in monobloc batteries like those shown in Figure 2-12.

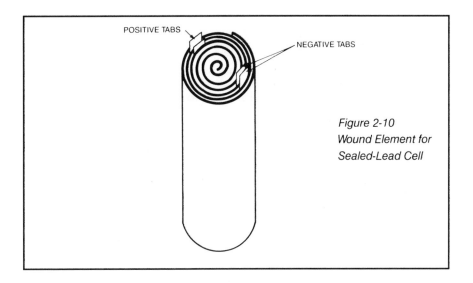

POSITIVE TABS

NEGATIVE TABS

*Figure 2-10
Wound Element for
Sealed-Lead Cell*

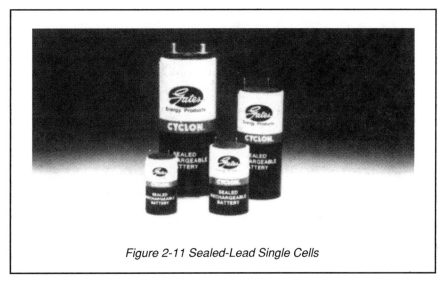

*Figure 2-11 Sealed-Lead Single Cells*

The single wound cell is a rugged, versatile unit that can be assembled into a wide variety of battery shapes and voltages.

As shown in Figure 2-13, assembly of the single cell begins with the wound element. This element is inserted into a plastic container. A plastic top containing a resealable pressure relief valve and provisions for electrical connections from the positive and negative plates is then sealed to the container. A metal can and plastic outer top provide additional mechanical protection to the cell. Just before final assembly, a carefully metered quantity of electrolyte is injected into the cell assembly. After cell assembly is complete, the cell receives its initial electrical charge. This formation charge transforms the paste on the positive and negative plates into the active materials (lead dioxide and sponge lead) required for proper cell function.

The Gates monobloc is an expansion of the single-cell product line designed primarily for high-volume OEM applications. Monoblocs may be available in 4 and 6-volt versions. These basic units are easily assembled in pairs to make 8 and 12 volt batteries as well.

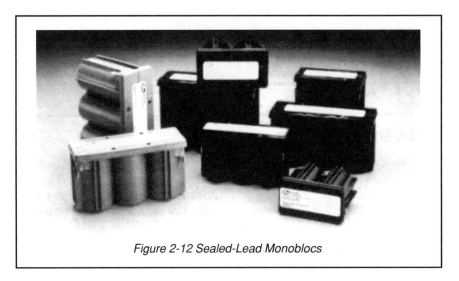

*Figure 2-12 Sealed-Lead Monoblocs*

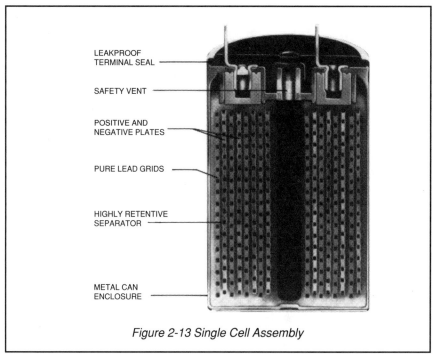

LEAKPROOF
TERMINAL SEAL

SAFETY VENT

POSITIVE AND
NEGATIVE PLATES

PURE LEAD GRIDS

HIGHLY RETENTIVE
SEPARATOR

METAL CAN
ENCLOSURE

*Figure 2-13 Single Cell Assembly*

Monobloc battery assembly (as shown in Figure 2-14) is somewhat different from single-cell assembly although it uses basically the same wound element construction. The elements are inserted into a plastic case that has individual containers for each element. Through-the-wall welds provide the electrical links between the cells. Positive and negative terminals are provided at opposite ends of the battery. The top contains a single resealable vent for all elements. Gas paths between the individual cell chambers permit recombination to function across cell boundaries, i.e. oxygen evolved at the positive in one cell may recombine at the negative of another cell.

It is important to note that the cell containers are electrically neutral for the sealed-lead battery. Electrical connections are made at discrete positive and negative terminals. This is different from the sealed nickel-cadmium battery where the can serves as the negative terminal while the top is the positive terminal.

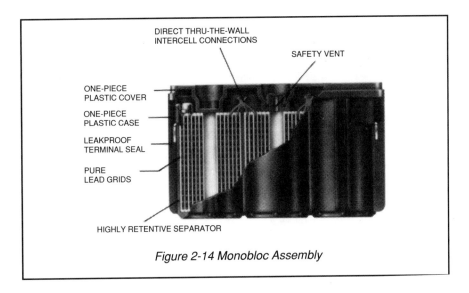

DIRECT THRU-THE-WALL
INTERCELL CONNECTIONS

SAFETY VENT

ONE-PIECE
PLASTIC COVER

ONE-PIECE
PLASTIC CASE

LEAKPROOF
TERMINAL SEAL

PURE
LEAD GRIDS

HIGHLY RETENTIVE SEPARATOR

*Figure 2-14 Monobloc Assembly*

## 2.5.3   OTHER FORMS OF LEAD BATTERIES

Wound sealed-lead batteries are the most advanced form of lead-acid battery available. They are particularly useful in applications where cleanliness, long-life and high energy density are important. However, different types of lead-acid batteries have existed for decades and continue to fill many needs. Other lead-acid batteries are most frequently classified as either flooded and gelled-electrolyte.

### 2.5.3.1   Flooded Lead-Acid

The flooded lead-acid battery today uses basically the design developed by Faure in 1881. It consists of a container with multiple plates immersed in a pool of dilute sulfuric acid. Recombination is minimal so water is consumed through the battery life and the batteries can emit corrosive and explosive gases when experiencing overcharge. So-called "maintenance-free" forms of flooded batteries provide excess electrolyte to accommodate water loss through a normal life cycle. Most industrial applications for flooded batteries are found in motive power, engine starting, and large system power backup. Today, other forms of battery have largely supplanted flooded batteries in small and medium capacity applications, but, in larger sizes flooded lead-acid batteries continue to dominate. By far, the biggest application for flooded batteries is starting, lighting, and ignition (SLI) service on automobiles and trucks. Large flooded lead-acid batteries also provide motive power for equipment ranging from forklifts to submarines and provide emergency power backup for many electrical applications, most notably the telecommunications network.

### 2.5.3.2   Gelled-Electrolyte

Since many of the problems with conventional flooded lead-acid batteries involve the liquid electrolyte, a variety of materials have been used in attempts to immobilize the liquid. This process ultimately came to fruition with the sealed-lead battery where essentially all the electrolyte is absorbed, either in the plate or the separator. Earlier work had used a gelling agent to produce a highly viscous electrolyte. These gelled-electrolyte batteries successfully overcame many objections to flooded batteries, especially in smaller sizes. Maintenance was eliminated and problems of corrosion and electrolyte leakage were greatly reduced from flooded batteries. These batteries have been surpassed in technology with the advent of the starved-electrolyte recombining sealed-lead cell using a fibrous separator.

# 2.6   Recent Developments in Rechargeable Batteries

As electrical and electronic products become increasingly essential to both consumers and business, the role of battery power in providing portable operation or backup for critical processes has become increasingly important. Although components with reduced power consumption and innovative power management strategies have made major contributions, much of the recent growth in new portable products has resulted from improvements in battery performance, especially rechargeable batteries. Further improvements in rechargeable technology, including both refinements in existing cell types and introduction of a radically different battery chemistries, will continue to offer opportunities to new product designers. At the same time, batteries of all forms, both primary and rechargeable, are coming under increasing scrutiny for their impact on the environment. Legislative and regulatory controls on battery disposal may influence the design of both portable and stationary products using batteries. All of these issues are very briefly discussed in the following sections; however, product designers are encouraged to consult with their battery suppliers to obtain the latest information on emerging new products and on current regulations concerning battery use and disposal.

## 2.6.1   ENHANCEMENTS IN CONVENTIONAL TECHNOLOGIES

The bulk of rechargeable portable power applications currently use nickel-cadmium cells. These cells have been the mainstays of portable power for the last 25 years. For much of their history, improvements in performance have been incremental. The $C_s$ (sub-C size) cell, which is the key size for many appliance and electronics applications, had increased 50 per cent in capacity over about 20 years leading up to 1988. In the next two years the capacity increased by another 50 per cent. These advanced-technology nickel-cadmium cells have obtained the increase in capacity by optimizing the use of the active materials within the cell rather than by major changes in cell chemistry. Increased-capacity nickel-cadmium cells have provided much of the foundation for the recent growth in portable computers, portable communications products, and appliances. For example the recent combination of advanced rechargeable cells, low-drain electronic components, and innovative energy management schemes now allow laptop computer users to spend most of a transcontinental flight working.

Continued gradual improvement in capacity for nickel-cadmium cells is projected to occur over the near-term as manufacturing process refinements are incorporated into production.

For those battery users for whom performance is the dominant requirement, there is a new type of rechargeable battery that extends the nickel-cadmium performance to new levels of capacity and power density.

## 2.6.2   NICKEL-METAL HYDRIDE CELLS

For those battery users who need high power in a small package and are willing to pay a premium for it, there is a new option. Battery manufacturers have begun to tap the potential of nickel-metal hydride couples which offer significant increases in cell power density. These extensions of the sealed nickel-cadmium cell technology to new chemistry are just becoming available for product applications such as notebook computers.

Some features of these new cells are discussed below:

### 2.6.2.1 Chemistry

The basis of the metal hydride technology is the ability of certain metallic alloys to absorb the smaller hydrogen atoms in the interstices between the larger metal atoms. Two general classes of materials have been identified as possessing the potential for absorbing large volumes of hydrogen: rare earth/nickel alloys generally based around $LaNi_5$ and alloys consisting primarily of titanium and zirconium. In both cases the some fraction of the base metals is often replaced with other metallic elements. The exact alloy composition can be tailored somewhat to accommodate the specific duty requirements of the final cell. Although cells using both types of materials are appearing on the market, precise alloy formulations are highly proprietary and specific to the manufacturer. These metal alloys are used to provide the active materials for the negative electrode in cells which otherwise are very similar to a nickel-cadmium cell. The metal hydride negative is used with a nickel hydroxide positive electrode and aqueous potassium hydroxide electrolyte as in the nickel-cadmium cell.

Use of a metal hydride provides the following reactions at the negative battery electrode:

*Charge*

When an electrical potential is applied to the cell, in the presence of the alloy, the water in the electrolyte is decomposed into hydrogen atoms which are absorbed into the alloy and hydroxyl ions as indicated below.

$$Alloy + H_2O + e^- \rightarrow Alloy[H] + OH^-$$

*Discharge*

During discharge, the reactions are reversed. The hydrogen is desorbed and combines with an hydroxyl ion to form water while also contributing an electron to the current.

$$Alloy[H] + OH^- \rightarrow Alloy + H_2O + e^-$$

This electrode has approximately 40 per cent more theoretical capacity than the cadmium negative electrode in a nickel-cadmium couple. The metal hydride negative works well with essentially the same nickel positive electrode used by conventional nickel-cadmium cells, but provides power densities that are 20 – 30 per cent higher in the finished cell.

The intrinsic voltage of the couple is approximately the same as with nickel-cadmium: 1.2 volts per cells. This helps in maintaining applications compatibility between existing nickel-cadmium cells and the new nickel-metal hydride cells.

### 2.6.2.2 Construction

The nickel-metal hydride couple lends itself to a wound construction similar to that used by present-day round nickel-cadmium cells. The basic components consist of the positive and negative electrodes insulated by plastic separators similar to those used in nickel-cadmium products. The sandwiched electrodes are wound together and inserted into a metallic can which is sealed after injection of a small amount of potassium hydroxide electrolyte solution. The result is a cell which bears a striking resemblance to current sealed nickel-cadmium cells. The nickel-metal hydride chemistry is also applicable to prismatic cell designs which evoke greater interest as product profiles become thinner.

### 2.6.2.3    Application Considerations

While details of application parameters are still being resolved, in general the nickel-metal hydride cell performs in a similar manner to conventional nickel-cadmium cells.

The cell volumetric energy density is currently between 25 and 30 per cent better than high-performance nickel-cadmium cells or close to double the energy density of "standard" nickel-cadmium cells. This translates to a present capacity for a Cs cell of 2700 milliampere-hours (mAh) in nickel-metal hydride vs. 2000 mAh for nickel-cadmium cells. Improved design and materials usage optimization is anticipated to result in about a 30 per cent increase in capacity for the metal hydride cell over the next five to six years. Although first development and application of the metal hydride cell has been for medium-rate applications, high-rate discharge capability has proven to be acceptable as well.

Behavior of the nickel-metal hydride cell on charge and its tolerance for overcharge is very similar to that of the nickel-cadmium cell. There are two areas where differences between the two chemistries should be recognized:

1) The metal-hydride charging reaction is exothermic while the nickel-cadmium reaction is endothermic. This means the metal hydride cell will warm as it charges whereas the nickel-cadmium cell temperature remains relatively constant until the cell moves into overcharge. However the metal hydride cell still retains the marked increase in temperature on overcharge which makes temperature-sensing charge termination schemes feasible.

2) The voltage profile on charge for nickel-metal hydride cells shows a less dramatic decline at overcharge than seen with nickel-cadmium cells. As nickel-cadmium cells move into overcharge, the voltage peaks and then begins a distinct decline. Some nickel-cadmium charging schemes have based charge termination on this negative slope to the voltage curve ($-\Delta v$ charging). Design of such systems to work with nickel-metal hydride cells will require careful attention to selection of the proper charge parameters.

Despite these differences, fast charging (return to full capacity in one hour or less) of the nickel-metal hydride cell with appropriate charge control and termination is fully feasible. In general, the same methods of charging used for nickel-cadmium cells work well for nickel-metal hydride although some of the control parameters may require adjustment for optimum performance and service life.

The nickel-metal hydride cell is expected to have a service life approaching that of the nickel-cadmium cell. Effects of operating temperature on both discharge performance and charge acceptance are also similar between cells.

One difference in performance between the two cell types occurs in self-discharge, i.e. loss of capacity in storage. The nickel-metal hydride cell currently self-discharges at a slightly higher rate than that seen with nickel-cadmium cells in storage. Development work is targeted at reducing the self-discharge rate to about the level seen with nickel-cadmium cells. Based on normal usage profiles expected with portable computers, self-discharge is not a major problem in this application.

Extensive testing of the environmental impact of the materials used in nickel-metal hydride cells has to date not identified any significant concerns. The occupational health and environmental disposal concerns associated with use of cadmium in nickel-cadmium cells appears to have been eliminated with nickel-metal hydride cells. Testing of metal-hydride cells indicates that their leachable cadmium content is below the EPA's threshold limit value. Cadmium content of metal-hydride cells is below the European Community's limit of 0.025 percent by weight. No other environmental problems have been associated with the metal hydride cell.

However, the performance and environmental benefits of nickel-metal hydride cells come at a price. At this time, the price per cell for metal hydride cells is approximately double an equivalent high-performance nickel-cadmium cell. This price premium is decline within the next couple of years as production volumes increase.

## 2.6.3  OTHER ADVANCED COUPLES

In addition to the nickel-metal hydride family of products, work continues to develop other couples that offer theoretical advantages, usually in energy density, over today's rechargeable batteries. Research efforts in rechargeable batteries have two major thrusts: 1) development of sealed batteries possessing higher energy densities for consumer/electronic applications, and 2) development of batteries for motive power applications such as electrically powered automobiles.

In the development of advanced sealed rechargeable batteries, the other option to nickel-metal hydride couples has been some form of rechargeable lithium-based couple. Rechargeable lithium batteries have shown much promise in offering a higher voltage (more than 2 volts for most forms) as well as higher energy density. Higher voltages minimize the number of cells required to reach common operating voltages with a consequent reduction in packaging issues. Many rechargeable cells using a lithium anode and a variety of cathode materials have been exhaustively researched, but to date none have demonstrated the cycle life and other performance characteristics necessary for successful commercial introduction.

## 2.6.4  ENVIRONMENTAL ISSUES

Essentially all batteries in widespread use today, rechargeable or not, have been associated with environmental, safety, or occupational health issues. While cadmium is usually identified as the greatest concern, improper disposal of batteries containing lead, mercury (often used in non-rechargeable batteries), and lithium also have environmental impacts. The nickel-metal hydride couple is virtually the only battery type that appears to be free of major environmental, occupational health, or safety concerns. To date, extensive research and testing performed on the various forms of nickel-metal hydride cells has not identified significant reasons to limit widespread use of the product.

Despite the concerns associated with current rechargeable batteries, most environmental organizations and many governmental agencies have promoted the replacement of primary batteries with rechargeables wherever feasible. Since one rechargeable cell can displace as many as 300 primary cells in some applications, this can have the effect of substantially reducing the quantity of potential pollutants entering the waste stream. However, the best approach to recycling or disposing of used batteries remains the subject of much debate.

To date there is no national plan in the United States for recycling batteries; the result has been a variety of state and local legislative and regulatory initiatives to control disposal of spent batteries. Outside the United States, regulatory controls on battery disposal are more advanced, but still vary widely from country to country in their severity and the level of implementation. Currently most attention is being placed on controlling primary batteries and small sealed rechargeable batteries such as found in consumer products; recycling programs are already in place and proving effective for larger battery types such as automotive batteries and industrial backup or motive power batteries. Although the proposed recycling/disposal plans for small cells vary widely, one common element seems to be a requirement that all batteries and battery packs be easily removable for replacement or disposal independent of the device they power. However, the controls on battery usage, which may even affect nickel-metal hydride cells, are changing rapidly. Designers of products containing batteries should consult with their battery suppliers regarding current requirements relating both to product design and to disposal of spent batteries.

# 2.7   Summary

Although vigorous effort has been expended to develop other forms of rechargeable batteries, nickel-cadmium and lead-acid batteries still represent the predominant commercially practical systems available today. In the super-premium area of spacecraft batteries nickel-hydrogen batteries have begun to displace nickel-cadmium batteries. Nickel-metal hydride batteries are beginning to find application in high-end commercial markets and hold the promise of ultimately replacing the standard nickel-cadmium cell.

Of the nickel-cadmium and lead-acid batteries available today, the sealed recombining systems offered discussed in this Handbook are the most advanced. They offer the best selection of performance features available in rechargeable batteries today. But, development work continues and further improvements in rechargeable batteries will continue to fuel the growth of the markets for battery-powered products.

*Section 3*

# Sealed Nickel-Cadmium Cells and Batteries

Sealed nickel-cadmium cells and batteries have proved over the last quarter century that they are a clean, reliable, lightweight source of power. These premium products are used in a diversity of consumer, industrial, and military devices needing battery power.

## FEATURES AND BENEFITS

Sealed nickel-cadmium batteries are especially well suited to applications where a self-contained power source increases the versatility or reliability of the end product. Among the significant advantages of sealed nickel-cadmium batteries which lead to their selection as the battery of choice for demanding applications are the following:

### High Energy Density

Nickel-cadmium batteries pack more energy into a given weight or volume than any other widely available rechargeable battery. Using nickel-cadmium batteries allows the design of small, lightweight devices such as instruments, transceivers, cordless telephones, video cameras and shavers.

### High-Rate Discharges

Nickel-cadmium batteries can deliver energy at very rapid rates. This high-rate discharge capability makes these batteries ideal for use in devices, such as power tools and appliances, that demand high power. In these applications a high-rate discharge capability leads to small, light, and economical battery-powered products. Nickel-cadmium cells can even supply repeated cycles with very high discharge rates.

### Fast Recharge

Some forms of sealed nickel-cadmium batteries are able to accept high charge rates for quick recharges. Although most nickel-cadmium battery applications have traditionally used a relatively low charge rate (which may require 16 to 20 hours or longer to achieve a full charge), special sealed nickel-cadmium batteries are now available that recharge in one hour or less.

For applications requiring a moderately fast recharge, quick-charge batteries can be completely recharged at the 0.3C rate in four to five hours, generally without the need for charge controls. They also tolerate overcharge at up to the 0.3C rate. These batteries have the advantage of requiring less complicated charging circuits than fast-charge batteries.

Fast-charge cells allow the user to recharge the battery in very short periods of time, normally within an hour. With this fast-charge capability, many cordless electric products may be used with little advance planning or preparation. Fast charging also permits the use of the full capacity of the battery several times a day, reducing the need for spares or larger batteries. However, the high currents required to obtain quick recharges must not be continued into overcharge. This requires special charger systems with controls to end the fast charging and switch to a maintenance rate when the battery is at full capacity.

## Long Operating Life

Nickel-cadmium batteries have an excellent operating life measured either by number of charge/discharge cycles or by years of useful life. Whether the battery is actively used through repetitive charging and discharging, or is simply kept on charge in a ready-to-serve condition, sealed nickel-cadmium cells offer a very long, trouble-free life. They will normally provide hundreds of charge/discharge cycles or operate for many years in a standby function. The long operating life gives nickel-cadmium batteries a life-cycle cost advantage over competing systems in many applications.

## Long Storage Life

The nickel-cadmium cell can be stored for extended periods of time at room temperatures without the need for maintenance charging. After extended storage, just one or two normal charge/discharge cycles usually results in rated performance. This is especially useful in applications where the battery spends long periods of time in the discharged state.

## Rugged Construction

Sealed nickel-cadmium batteries are very rugged devices, both physically and electrochemically. They have excellent resistance to shock and vibration. The nickel-cadmium battery lends itself to applications involving temperature extremes, tough physical use, or other demanding requirements.

## Operation Over a Broad Range of Temperatures

Nickel-cadmium batteries perform well over a wide range of operating temperatures. Standard sealed nickel-cadmium batteries deliver usable capacity in applications where temperatures may drop to –40°C or rise to 50°C. Special high temperature cells are available for applications in which the battery operates continuously between 50°C and 70°C.

## Operation in a Wide Range of Environments

Sealed nickel-cadmium cells function in a full vacuum or in a positive pressure environment. They are also generally insensitive to variations in relative humidity.

## Operation in Any Orientation

Sealed cells may be mounted and operated in any position or attitude. Product designers now have the freedom to locate the battery in the best position for the product.

## Maintenance-Free Use

Nickel-cadmium sealed-cell batteries are an excellent choice for applications in which periodic maintenance would be difficult or costly. Since the gases generated in normal operation recombine within the cell, sealed cells survive repeated normal duty cycles with no loss of active material or electrolyte. The sealed nickel-cadmium cell is an install-and-forget power source.

## Continuous Overcharge Capability

Sealed nickel-cadmium batteries accommodate extended charging at recommended charge rates, with no noticeable effect on performance or life. This means that simple and inexpensive chargers may be used successfully with these cells. Because of their ability to handle continuous overcharge, sealed nickel-cadmium batteries are widely used in standby power applications and in appliances that maintain a readiness to serve.

## Discharge Voltage Regulation

Sealed nickel-cadmium batteries have a nearly flat voltage profile throughout most of the discharge. As a result, battery-powered equipment can be designed to operate in a narrow voltage band rather than having to accommodate the wide voltage spreads typical of many other batteries.

## APPLICATION EXAMPLES

With the attributes described above, sealed nickel-cadmium batteries have found uses in a wide array of equipment. Some examples include:

**Portable Power**
- Portable Lighting
- Cordless Tools and Appliances
- Toys and Hobbies
- Portable Electronics
- Portable Communications

**Alternative Power**
- Computers
- Consumer Electronics
- Instrumentation

**Standby Power**
- Emergency Lighting
- Security Alarm Systems
- Electronics Backup

## SECTION CONTENTS

The remainder of this section provides the information needed to properly select the correct sealed nickel-cadmium battery for an application and then design the battery into the system. It begins with a discussion of discharge performance in Section 3.1 since this information is vital to selecting the proper battery for the application. To discharge properly, the battery must be charged correctly. Incorrect charging is also a major cause of reduced battery life. Suggestions for tailoring charging to the application are provided in Section 3.2. Section 3.3 explains how to store sealed nickel-cadmium batteries and cells. Understanding the tradeoffs that affect battery life is the theme of Section 3.4. Section 3.5 provides a variety of application information on sealed nickel-cadmium cells and batteries. This includes a discussion on comparing the economic benefits of batteries, information on packaging and mounting batteries, operating environment considerations in using batteries, and a brief review of some typical applications for sealed nickel-cadmium cells. Since testing is often the only way to understand how a battery will function in a specific application, Section 3.6 describes some approaches to battery testing. Section 3.6 also includes information on quality control and on standardized battery specifications. Finally, Section 3.7 covers the safety precautions that apply to sealed nickel-cadmium batteries and cells.

# 3.1 Discharge Characteristics

The nickel-cadmium cell stores electrical energy and makes that energy available during discharge. The cell's behavior during discharge is central to determining its suitability for any application. This section qualitatively describes the discharge characteristics common to many sealed nickel-cadmium cells. Using this information along with quantitative data for specific cell types (such as that presented in Appendix A), a product designer can evaluate whether a specific cell will meet the discharge requirements of a particular application.

## 3.1.1 GENERAL

The discharge parameters of concern are cell (or battery) voltage and capacity (the integral of current multiplied by time). The values of these two discharge parameters are functions of a number of application-related factors as described in this section. The general shape of the discharge curve, voltage as a function of capacity (or time if the current is uniform), is shown in Figure 3-1. The discharge voltage of the sealed nickel-cadmium cell typically remains relatively constant until most of its capacity is discharged. It then drops off rather sharply. The area of relatively constant voltage is called the *voltage plateau*. The flatness and the length of this plateau relative to the length of the discharge are major features of sealed nickel-cadmium cells and batteries.

The discharge curve, when scaled by considering the effects of all the application variables, provides a complete description of the output of a battery. Differences in design, internal construction, and conditions of actual use of the cell affect the performance characteristics. For example, Figure 3-2 illustrates the average (typical) effect of discharge rate.

The remainder of this chapter will define the discharge curve in terms which will allow the construction of a complete discharge curve ( voltage vs. both capacity and run time) for any cell (battery) proposed for an application using variables and parameter values appropriate to that application.

### 3.1.1.1 Cell Discharge Performance Measures

The construction of a discharge voltage curve requires that the terms of reference be defined. These are developed for a geometric shape in Figure 3-3 and then related to the typical discharge voltage curve in Figure 3-4.

Dimensions used to describe the geometric shape in Figure 3-3 are the values of A and B. With the addition of a value for $\Delta Y$, the figure is completely dimensioned by the three values. If $\Delta Y$ were always a fixed value, or even some fixed percentage of B, only the two values A and B would be necessary to convey complete dimensional information about the figure.

Figure 3-4 illustrates the general shape of the discharge voltage curve for sealed nickel-cadmium cells. It bears a strong similarity in shape to Figure 3-3. The X axis in the previous figure becomes capacity (or time) and the Y axis becomes cell voltage.

Just as the value of B is the average value of height of the area in Figure 3-3, the value of mid-point voltage (MPV) is, approximately, the average value of discharge voltage for a cell. Just as A multiplied by B is the area of Figure 3-3, MPV multiplied by cell capacity is the approximate area of Figure 3-4. The area under the

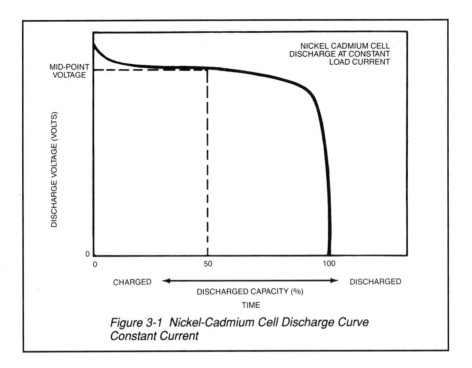

*Figure 3-1  Nickel-Cadmium Cell Discharge Curve Constant Current*

curve in Figure 3-4 is the energy supplied by the cell. The three parameters of MPV, cell capacity, and $\Delta V$ dimension describe, approximately, the entire discharge curve. By defining how these three parameters vary with discharge conditions, a close approximation of the actual discharge curve may be drawn. This procedure is described in Sections 3.1.2, 3.1.3 and 3.1.4.

### 3.1.1.2    Cell Capacity Defined

Battery or cell capacity simply means an integral of current over a defined period of time.

$$\text{Capacity} = \int_{\Delta t} i \, dt$$

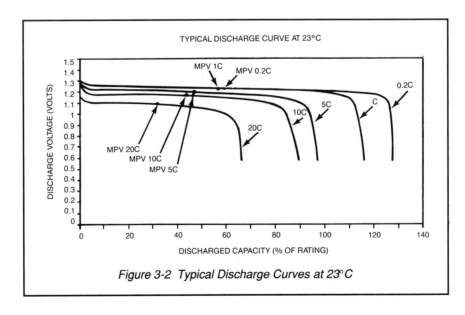

*Figure 3-2  Typical Discharge Curves at 23°C*

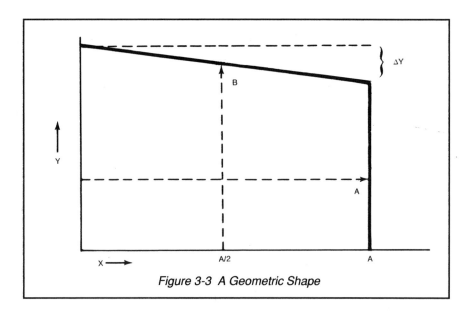

*Figure 3-3 A Geometric Shape*

This equation applies to either charge or discharge, i.e. capacity added or capacity removed from a battery or cell. Although the basic definition is simple, many different forms of capacity are used in the battery industry. The distinctions between them reflect differences in the conditions under which the capacity is measured. Commonly used forms of capacity are introduced in Table 3-1 and then discussed in Section 3.1.3.

## 3.1.2    CELL DISCHARGE VOLTAGE PERFORMANCE

This section describes the voltage (vertical or Y) dimension of the discharge curve. Section 3.1.3 will deal with the capacity (horizontal or X) dimension. This section also describes the application conditions that influence voltage delivery.

| | |
|---|---|
| Standard Conditions | = Laboratory Conditions: charge/rest/discharge rates/voltage/temperature |
| Standard Capacity | = Cell capacity measured under standard conditions. |
| Rated Capacity | = The minimum standard capacity. |
| Actual Capacity | = Capacity of a fully charged cell measured under non-standard conditions except standard end of discharge voltage (EODV). |
| Retained Capacity | = Capacity remaining after a rest period. |
| Available Capacity | = Capacity delivered to a non-standard EODV. |
| Dischargeable Capacity | = Capacity which a cell can deliver before it becomes fully discharged. |

*Table 3-1  Capacity Terminology Defined*

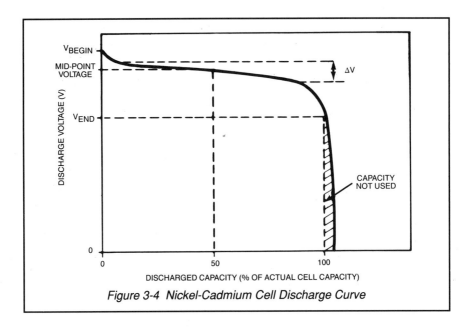

*Figure 3-4 Nickel-Cadmium Cell Discharge Curve*

### 3.1.2.1   Mid-Point Voltage

Figure 3-4 defines the mid-point voltage (MPV) as the voltage delivered by the cell when half of the total actual cell capacity has been discharged. MPV is also a reasonable approximation of the average voltage delivered during an entire constant-current discharge. Using this approximation, the energy delivered during discharge can be calculated just as if the discharge curve were a rectangle. Energy delivered is simply MPV multiplied by cell capacity with the results expressed in watt-hours or watt-minutes.

The remainder of Section 3.1.2 describes the methods for estimating discharge MPV as a function of application conditions.

### 3.1.2.2   Cell Discharge Equivalent Circuit

Figure 3-5 represents a Thévenin equivalent discharge circuit, which can be used to visualize and understand the performance of the nickel-cadmium cell during a typical discharge. Using the equivalent circuit is a simple means of estimating the actual voltage delivered by the cell during discharge.

The cell's equivalent voltage and resistance are not constant at their standard condition values but are functions of a variety of application variables such as state of discharge, cell temperature, and cell use history. Discussions in Sections 3.1.2.3 and 3.1.2.4 will describe how these application variables influence the values of effective cell no-load voltage ($E_o$) and effective cell internal resistance ($R_e$) and through them the actual discharge terminal voltage delivered by the cell.

The average cell discharge voltage, illustrated in Figure 3-4, can be defined using the equivalent circuit described above. The extended term, or steady-state, discharge voltage was shown to be:

$$E = E_o - IR_e$$

At the midpoint of the discharge curve, where $E_o = MPV_o$, this equation becomes:

$$MPV = MPV_o - IR_e$$

The equivalent circuit is presented only to help in explaining the voltage delivery behavior. The electrical values of effective no-load voltage ($E_o$), effective cell inter-

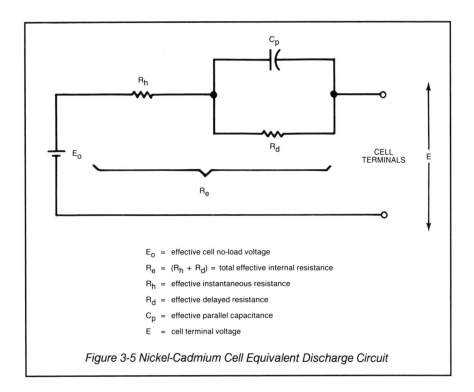

$E_o$  = effective cell no-load voltage

$R_e$  = $(R_h + R_d)$ = total effective internal resistance

$R_h$  = effective instantaneous resistance

$R_d$  = effective delayed resistance

$C_p$  = effective parallel capacitance

E   = cell terminal voltage

*Figure 3-5 Nickel-Cadmium Cell Equivalent Discharge Circuit*

nal resistance ($R_e$), and effective capacitance ($C_p$) are used, not necessarily because they are physically present in the cell as lumped values, but because the cell performs approximately as though they are.

### 3.1.2.2.1 *Steady-State Performance*

The effective steady-state internal cell resistance $R_e$ consists of two resistive elements in series, $R_h$ and $R_d$. When the cell begins to discharge, its voltage immediately falls from its no-load value to a value determined by $R_h$. During the first few seconds of discharge, the delivered cell voltage drops rapidly from that initial level determined only by $R_h$ to the extended time steady-state discharge level set by $(R_h + R_d)$ which is the total effective internal resistance $R_e$. Figure 3-6 shows this behavior. The practical result of using the equivalent circuit approach, with its values of $E_o$ and $R_e$, is illustrated in Figure 3-7. On this chart the discharge terminal voltage response of a representative cell is plotted as a function of load discharge current. The equation of this loadline is ($E = E_o - IR_e$) for the normal case of extended term, steady-state, discharge.

### 3.1.2.2.2 *Dynamic Transient Response*

Because of the relatively short duration of the dynamic transient response it can simply be ignored in most applications. When the load switch is closed the cell voltage response begins on the upper load regulation line of Figure 3-7 and then transfers exponentially (time constant = $C_p \times R_d$) to the steady-state load regulation line. This transfer occurs as the effective capacitor is charged.

The reverse effect occurs when the load is suddenly reduced or terminated. The equation of the load line for a discharge pulse of very short duration is thus ( $E = E_o - IR_h$). Typical values for $R_h$ as a fraction of $R_e$ range from 25 to 75 per cent, depending primarily on the design of the cell.

The combination of the parallel portion of the resistance ($R_d$) and the effective

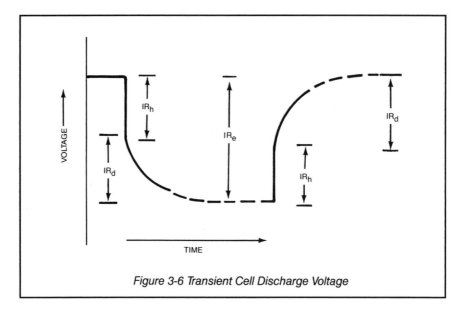

*Figure 3-6 Transient Cell Discharge Voltage*

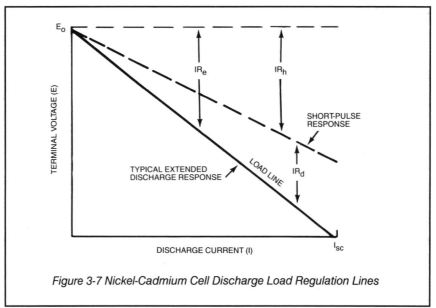

*Figure 3-7 Nickel-Cadmium Cell Discharge Load Regulation Lines*

capacitance ($C_p$) has a time constant of approximately one to three per cent of the cell discharge time, varying somewhat with cell temperature and discharge rate. Typical values for this time constant become shorter as the discharge rate increases and longer as the cell temperature decreases.

For very short discharge periods (pulses), $C_p \times R_d$ can almost be ignored. The effective internal resistance, considered to consist only of $R_h$, causes an instantaneous drop in discharge voltage when the load switch is closed and an instantaneous rise in voltage when the load switch is opened. These effects were illustrated in Figure 3-6. For repetitive discharge pulses, the amount of average voltage drop depends approximately on the steady-state resistance ($R_e$) multiplied by the average current. The average current is the current during the *ON* pulses multiplied by the duty cycle. The actual voltage oscillates above and below this average value.

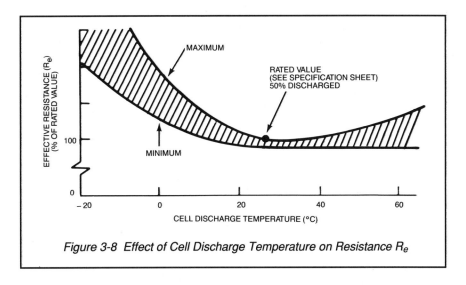

*Figure 3-8  Effect of Cell Discharge Temperature on Resistance $R_e$*

### 3.1.2.3    Variables Influencing $R_e$

The value of $R_e$, for any particular cell, is influenced by the value of several application variables. The most important of these are cell temperature during discharge, the amount of capacity already discharged, and the history (age and previous use) of the cell.

#### 3.1.2.3.1  $R_e$ as a Function of Cell Discharge Temperature
The effective internal resistance, ($R_e$), changes with cell temperature as indicated in Figure 3-8. The resistance is at a minimum at cell temperatures between 20°C and 40°C. Below 20°C, $R_e$ begins to rise, primarily because of the difficulty of ionic flow through the increasingly viscous electrolyte of the cell.

#### 3.1.2.3.2  $R_e$ as a Function of State of Discharge
Effective internal resistance also depends on the degree to which the cell has been discharged as shown in Figure 3-9. The horizontal axis represents the capacity

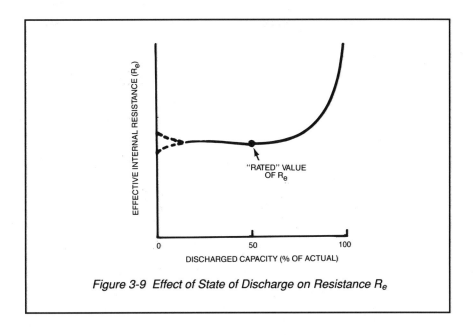

*Figure 3-9  Effect of State of Discharge on Resistance $R_e$*

removed from the cell as a percentage of the total actual capacity that the cell could deliver under the prevailing operating conditions and environment. $R_e$ remains a rather stable value for the major portion of the total discharge period. For a short period at the beginning of the discharge, the effective resistance is quite difficult to measure accurately and thus is indicated as being somewhat variable.

Since the variations of $E_o$ during discharge are approximately linear and also relatively small, the typical nickel-cadmium discharge voltage curve shape (Figure 3-4) is almost a mirror image of the curve of $R_e$ as a function of discharged capacity shown in Figure 3-9. The discharge voltage of nickel-cadmium cells remains relatively flat, subject primarily to a slowly decreasing value of $E_o$, throughout most of the discharge period. Toward the end of discharge the significant increase in the effective internal cell resistance $(R_e)$ is the principal cause of the steep decrease in discharge voltage. This increase in $R_e$ is the result of having converted most of the available active material to the lower valence state, thus making the ionic conduction path to the remaining unconverted material more and more tortuous.

The value of $R_e$ near the midpoint, 50 per cent on Figure 3-9, is relatively constant and therefore insensitive to errors in locating the midpoint. This insensitivity allows design of a $R_e$ measurement procedure that reduces measurement errors.

### 3.1.2.3.3 $R_e$ as a Function of History of Use and Cell Design

The $R_e$ characteristic of a cell also depends on how the cell has been used, its history. The principal effect of cell history is a very slowly increasing value of $R_e$ as a result of both use and abuse.

A cell's effective internal resistance $(R_e)$, holding all other factors constant, is approximately inversely proportional to its total designed capacity. For instance, two identical cells discharged in parallel have twice the capacity and one-half the effective internal resistance of each individual cell.

### 3.1.2.3.4 Calculation and Measurement of $R_e$

The effective internal resistance $(R_e)$ of a cell or battery is the relationship between steady-state current and delivered voltage at a specific point in the discharge such as the mid-point (50 per cent of the capacity removed). For example, consider a specific cell. The relationship between two values of MPV (50 per cent discharged) at two different constant-current discharge rates determines the cell's $R_e$ by the equation:

$$R_e = -(MPV_2 - MPV_1)/(I_2 - I_1)$$

This relationship extends to:

$$R_e = -\Delta E/\Delta I$$

or:

$$R_e = -(E_2 - E_1)/(I_2 - I_1)$$

to relate the change in voltage delivered by a cell to the difference in discharge rate when the discharge rate is changed. This approximate relationship exists only for changes from one steady state to another, allowing time for dissipation of the transient.

In the ANSI Standard C18, 2-1984: *Specifications for Sealed Rechargeable Nickel-Cadmium Cylindrical Bare Cells*, paragraph 9.4 establishes a standard procedure for making this type of internal resistance measurement. This procedure is used by many, but not all, cell manufacturers. Some report only an AC measurement of internal impedance which can be read directly from commercially available milli-ohmmeters. This AC internal impedance is substantially lower than that determined

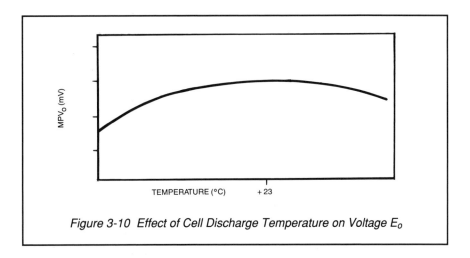

*Figure 3-10 Effect of Cell Discharge Temperature on Voltage $E_o$*

by discharge readings. While it may be useful as a cell diagnostic tool, it gives erroneous results when used to calculate voltage drop or other cell performance parameters.

### 3.1.2.4  Variables Influencing $E_o$

In addition to the effect of $R_e$ and the discharge rate, other application conditions affect the discharge voltage by changing the magnitude of the effective no-load voltage $E_o$.

#### 3.1.2.4.1 $E_o$ as a Function of Cell Discharge Temperature

The cell no-load voltage is near its peak at room temperature as shown in Figure 3-10. It declines as the temperature increases or decreases. However the decline is much more pronounced as the cell cools. While $E_o$ at the midpoint of the discharge curve (MPV$_o$) changes only slightly at cell temperatures above 20°C; the value of MPV$_o$ decreases significantly as temperatures drop below room temperature. The total decrease in discharge voltage at low cell temperatures is therefore the result of this change in MPV$_o$ combined with the significant increases in the value of $R_e$ at low cell temperature (shown in Figure 3-8).

Note that the temperature effect on no-load voltage is substantially different from the temperature effect on overcharge voltage discussed in the charging section (Section 3.2.4).

#### 3.1.2.4.2 $E_o$ as a Function of the State of Discharge

The no-load cell voltage also varies moderately during the discharge process as a function of the capacity removed from the cell.

Figure 3-11 illustrates the effect that the degree of discharge of a cell has on $E_o$. It plots the typical effective no-load voltage of a sealed nickel-cadmium cell at 23°C versus the capacity discharged from the cell. The effective no-load voltage remains relatively flat throughout most of the discharge period. There is an approximately linear 100 mV drop from fully charged to fully discharged. The dashed line (A — B) in the figure describes this gradual decrease in $E_o$, and is the $\Delta V$ effect in Figure 3-4.

In general, the no-load voltage effects have comparatively little impact on the cell discharge voltage. The effective internal resistance of the cell ($R_e$) has a more significant influence on discharge voltage. Resistance effects are especially important as the end of a discharge is approached and in high-rate discharge applications.

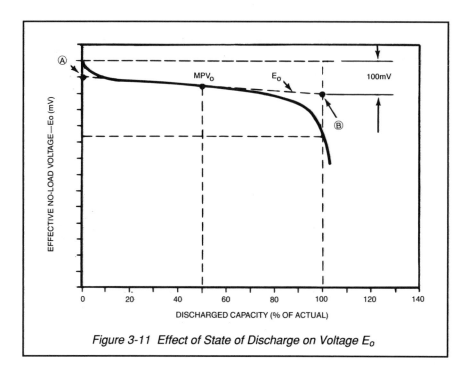

*Figure 3-11  Effect of State of Discharge on Voltage $E_o$*

### 3.1.2.5    Summary of Voltage Effects

For a review of the various terms and effects on the voltage delivered by a cell, refer to the exaggerated curve of Figure 3-12 and the accompanying equations. Voltage delivery is a function of discharge current and two cell variables — the equivalent no-load voltage ($E_o$) and the effective internal resistance ($R_e$). The curve of Figure 3-12 shows the effect of state of discharge on both $E_o$ and $R_e$.

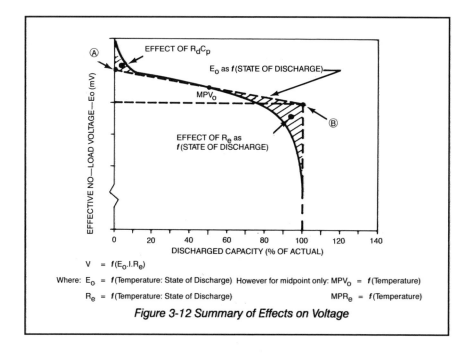

$$V = f(E_o . I . R_e)$$

Where: $E_o$ = $f$(Temperature: State of Discharge) However for midpoint only: $MPV_o$ = $f$(Temperature)

$R_e$ = $f$(Temperature: State of Discharge)          $MPR_e$ = $f$(Temperature)

*Figure 3-12 Summary of Effects on Voltage*

The principal load regulation voltage loss is a function only of discharge rate and $R_e$ as was described in Figure 3-7. $R_e$ depends on cell design and cell discharge temperature, as in Figure 3-8.

The average no-load supply voltage $MPV_o$ is relatively independent of operating variables other than the effect of low cell temperature shown in Figure 3-10. Except for the temperature effect on $MPV_o$, $E_o$ depends only on discharged capacity. This relationship is relatively independent of other operating variables. The state-of-discharge effect may be approximated as a nearly linear $\pm 50$ mV relative to $MPV_o$ over the range of actual capacity of the cell.

The two deviations, shown as shaded areas, from the straight line (A — B) of Figure 3-12, are at the beginning and end of the discharge. The transient result of $R_d$ and $C_p$ at the beginning of the curve keeps the voltage above the steady-state predicted value. Also contributing to the initial transient is the depletion of electrode surface charge which occurs at the beginning of discharge of a fully charged cell. Some valence shift may also be included. The effect of $R_d$ and $C_p$ reappears, of course, if the load is removed and then reapplied. The rapidly increasing $R_e$ as a function of the depleting state of charge at the end of the discharge curve causes battery performance to drop below the linear approximation.

Other than these two deviations from the straight line, discharge rate and discharge temperature are the only application parameters having significant effects on the discharge voltage. Both of these effects may be conveniently calculated, as they will be in Section 3.1.4.

## 3.1.3   CELL DISCHARGE CAPACITY PERFORMANCE

This section describes the capacity (the horizontal or X) dimension of the discharge curve (Figures 3-3 and 3-4). Having previously considered the effect of application conditions on the cell's voltage delivery, it is now time to consider application effects on the capacity that may be extracted from the cell.

The capacity delivered by a cell is the integral of current (electron flow) over time or basically the gross number of electrons supplied by the cell to the outside circuit. The number of electrons that the cell will supply is a function of both how the cell is used (current flow, charge method, duty cycle) and the environment, i.e. operating temperature. This section will first refine the various definitions of capacity and then describe some of the parameters that affect the capacity a cell will deliver.

### 3.1.3.1   Cell Capacity Definitions and Ratings

There are a variety of possible meanings of capacity depending on the conditions under which it is measured. Cell capacity indicates the amount of charge (electrons) that the fully charged cell is capable of storing and then delivering to the electrical load. It is normally expressed in units of ampere-hours although milliampere-hours, ampere-minutes, or other units of current multiplied by time may be used. It can be conveniently related to volt-ampere-minutes, watt-hours, or some similar measurement of energy delivery by multiplying the ampere-hours by the average delivered voltage (MPV).

*Standard cell capacity* measures the total capacity that a relatively new production cell can store and discharge under a defined standard set of application conditions. It assumes that the cell is fully charged at standard cell temperature at the cell specification rate, and that it is discharged at the same standard cell temperature at a specified standard discharge rate to a standard end-of-discharge voltage (EODV). The value of standard cell capacity may lie anywhere within the statistical distribu-

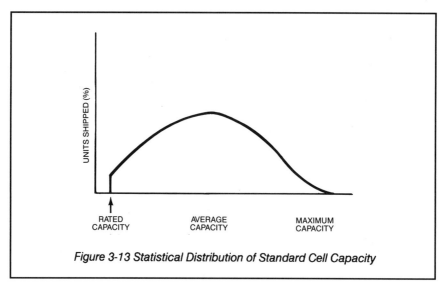

*Figure 3-13 Statistical Distribution of Standard Cell Capacity*

| Conditions | Charging | | | Resting | | Discharging | | |
|---|---|---|---|---|---|---|---|---|
| Capacity Terminology | Cell Temp | Rate | Amount (4) | Ambient Temp | Time | Cell Temp | Rate | EODV (3) |
| Standard Conditions | 23°C | 0.1C | 20 hr Min. | 23°C | Minimum | 23°C | 1C | 0.9V |
| *Standard* Cell Capacity (1) | 23°C | 0.1C | 20 hr Min. | 23°C | Minimum | 23°C | 1C | 0.9V |
| *Rated* Cell Capacity (2) | 23°C | 0.1C | 20 hr Min. | 23°C | Minimum | 23°G | 1C | 0.9V |
| *Actual* Cell Capacity | @ | @ | Fully Charged | @ | Minimum | @ | @ | 0.75MPV (3) |
| *Retained* Cell Capacity | @ | @ | Fully Charged | @ | @ | @ | @ | 0.75MPV (3) |
| *Available* Cell Capacity | @ | @ | Fully Charged | @ | Minimum | @ | @ | @ |
| *Dischargeable Capacity* | @ | @ | @ | @ | Minimum | 23°C | 1C | 0.9V |

*Battery Capacity* is approximately equal to the capacity of the minium capacity cell in the battery when that cell is discharged to zero volts.

Notes: (1) - *Standard* Cell Capacity = *Actual* Capacity of a particular production cell under Standard conditions. Characteristically ranges from 100 to 130% of *Rated* Capacity.

(2) - *Rated* Cell Capacity = Minimum *Standard* Cell Capacity = C
See the *Gates Energy Products* Specification sheet for the value of C assigned to each cell design.

(3) - EODV = End of Discharge Voltage
MPV = Mid Point Voltage

(4) - The amount of charge required for the measurement of Cell Capacity, is *Full Charge* into overcharge.
2.0C Ahrs (20 hours at the 0.1C rate) at 23°C, is required to insure that a cell with maximum *Standard* Cell Capacity is fully charged.
2.4C Ahrs (24 hours at 0.1C rate) at 23°C, is required on initial charge or after extended storage —

( @ ) = Any specified value of a test condition other than a Standard value.

*Table 3-2 Capacity Terms Defined*

tion of capacity as manufactured. Figure 3-13 illustrates a typical capacity distribution. *Unless otherwise stated, this Handbook uses standard conditions.*

When any of the application conditions differ from standard, the capacity of the cell may change. A new term, *actual cell capacity*, is used for all nonstandard conditions that alter the amount of capacity which the fully charged new cell is capable of delivering when fully discharged to a standard EODV. Examples of such situations might include subjecting the cell to a cold discharge or a high-rate discharge.

That portion of actual cell capacity which can be delivered by the fully charged new cell to some nonstandard end-of-discharge voltage is called *available cell capacity*. Thus if the standard EODV is 0.9 volts per cell, the available capacity to an end-of-discharge voltage of 1.0 volts per cell would be less than the actual capacity.

Cells are rated at standard specified values of discharge rate and other application conditions. *Rated cell capacity* (**C**) for each cell type is defined as the minimum standard capacity to be expected from any cell of that type when new. The rated value must also be accompanied by the hour-rate of discharge upon which the rating is based (e.g. 1 hr, 5 hr, 10 hr, 20 hr, etc).

Rated cell capacity is always a single specific designated value for each cell model (type, size and design), as contrasted with the statistically distributed values for all other defined capacities. Thus a group of $C_s$ cells with a rated capacity of 1.4 amp-hours might have standard capacities ranging from 1.4 to 1.7 amp-hrs with an average of 1.55. The Gates process for the manufacture of nickel-cadmium cells produces a comparatively tight spread in the overall distribution of standard capacity as shown in Figure 3-13.

Figure 3-13 refers to single-cell capacity and NOT multi-cell battery capacity. In any multicell battery, the lowest capacity cell in the battery determines its capacity. The distribution of battery capacity, therefore, has the same minimum value as in Figure 3-13 (rated capacity), but its maximum capacity may be somewhat reduced. This reduction depends on the number of cells in the battery and the width (statistical variance) of the capacity distribution of the particular population of cells from which the batteries are actually constructed. For example, if only identical capacity cells were used within each battery, the distribution of battery capacity would be the same as the distribution of cell capacity. For further discussion of the effects of cell capacity distribution on battery capacity, see Section 3.1.5.

If a new battery is stored for a period of time following a full charge, some of its charge will dissipate. The capacity which remains that can be discharged is called *retained capacity*. (See Section 3.1.3.4 for explanation and discussion of this term.)

Table 3-2 summarizes the specific conditions for each of the capacities defined above.

### 3.1.3.2 Measurement of Fully Charged Capacity

The capacity of a cell, or battery, is normally measured by completely discharging it while integrating the current over the period of the discharge. Variations in the capacity measurement procedure can result in data inconsistencies.

The most common method of measuring capacity is to discharge the cell with a constant-current load. The load circuit adjusts to maintain a constant discharge current as the cell voltage declines. Recording cell voltage versus time results in a discharge curve similar to Figure 3-4. Calculation of discharged cell capacity is thus only a multiplication of the time needed to reach the specified end-of-discharge voltage (EODV) times the current. An added refinement is the simultaneous use of a current integrator or current regulator to ensure that the load is stable and accurate in maintaining the constant current. A variety of packaged loads designed specifically for constant-current discharges are available for capacity measurement.

An older, less common, and less accurate method of measuring capacity is to place a fixed resistance load across the cell terminals and monitor the voltage as a function of time as the cell discharges. With a fixed resistance, the current decreases as the cell voltage declines. A recorder is used to record this voltage. The discharge recording of variable voltage is translated to current and then manually integrated over time to calculate the discharged capacity. Use of a current integrator in the circuit can speed the capacity measurement. Unfortunately, the discharge current, which influences actual cell capacity, is variable in this procedure. Thus, relating the results to other application conditions can be quite difficult.

Measured cell capacity depends also on the end-of-discharge voltage used in the measurement. For most accurate results in measuring total battery capacity, the voltage used to terminate the discharge should be below the knee of the discharge curve. This simply means that the end of the discharge should occur after the cell has left the flat plateau of the discharge curve and the voltage is falling rapidly. An EODV of 0.9 V for 1C rated discharge is typically used.

Higher values of EODV, when used in measurement procedures, may increase the inaccuracy of the results. For EODV's on the voltage plateau, for example, voltage is dropping slowly with time so small errors in measured voltage may result in significant errors in the time (capacity) to the end of the discharge. Once the cell is off the plateau, the voltage falls very rapidly and the remaining effective capacity is slight so there is little capacity difference between an EODV of 0.9 volts per cell and one of 0.6.

Definitions of standard 1C rate cell capacity used in this Handbook are based on a discharge at a constant current to a cell EODV of 0.9V (75 per cent of 1.2V).

Some off-standard conditions, such as high discharge rates or low cell temperatures, will significantly reduce the average voltage delivered by the cell. Because of this, the cutoff voltage (EODV) is defined to be 75 per cent of MPV. This relates the minimum useful voltage for any load to the average voltage (MPV) being used by that load and for which the load device is designed. Cell and battery cutoff voltages are discussed in more detail in Section 3.1.5.4.

### 3.1.3.3    Actual Cell Capacity at Off-Standard Conditions

Standard cell capacity is the capacity of a particular individual cell under standard conditions of charge, rest, and discharge. Standard cell capacity typically ranges from 100 to 130 per cent of the rated (minimum) value. Actual cell capacity is the capacity of that same individual cell when the conditions under which the cell is charged and discharged differ from standard conditions. The difference between standard and actual capacity normally results from variations in cell temperature during charge, charge rate, cell temperature during discharge, or discharge rate.

The effects of each of these off-standard conditions are defined by the amount of derating of standard cell capacity needed to produce the actual cell capacity for those conditions. The following sections discuss in general terms how the derating factors are developed. The actual derating curves for examples of various cell types are provided in Appendix A. Section 3.1.4 then presents a format for this information that allows its use to estimate discharge curves for specific applications.

First, the baseline for derating must be established. The most conservative baseline possible for standard capacity is the rated minimum capacity. Either that value or the more generous typical or average standard capacities may be used as the basis for the derating procedure for actual capacity. The key is to be consistent and to understand the differences in the result produced.

### 3.1.3.3.1 *Charging Conditions*

Both cell temperature during charge and the amount of charging (capacity input) have an impact on the capacity that the cell is capable of discharging. These differences can be defined by using the following variables:

- Charge Temperature ( °C)
- Charge Rate (A)
- Charge Time (hr)

*Charge Temperature* — Elevating the charge temperature of the cell influences the charge behavior in two ways, as discussed in Section 3.2.2.1. First, a higher cell temperature during charge requires significantly more charge input for the cell to reach full charge and achieve its actual capacity at that elevated cell temperature. Second, in addition to the increased charging needed to attain full charge, the in-

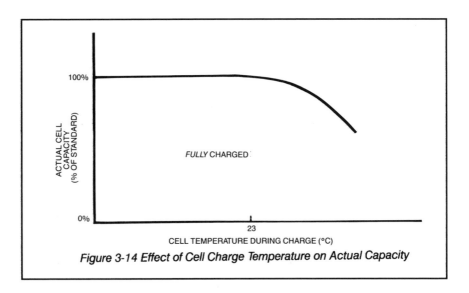

*Figure 3-14 Effect of Cell Charge Temperature on Actual Capacity*

creased cell temperature reduces the actual cell capacity below the standard capacity value. This is true even when the cell is fully charged at those elevated cell temperatures — for example 72 hours or more at C/10-rate at 48°C. Assuming that the cell is fully charged for measurement purposes, Figure 3-14 shows the effect of cell temperature during charge on actual capacity.

Although actual cell capacity is not affected by charge temperatures below 23°C, charge currents must be decreased below the standard rated value when operating standard and quick-charge cells at cell temperatures below 0 and 10°C respectively. See Figure 3-63 or the individual cell specification sheet for the allowable value of charge current, as a function of cell temperature, which keeps the cell pressure low enough to prevent venting while still reliably charging the cell.

*Charge Rate* — The effect of charge rate (standard and quick) on actual cell capacity is relatively negligible as long as the cell reaches overcharge and the charge rate is within the cell specifications. Charge rates below the minimum cell specification rate may lower the actual cell capacity even if the cell reaches overcharge. On the other hand, fast-charge rates, although they must be limited in duration to prevent cell damage, may produce actual cell capacities which are modestly greater than 100 per cent of standard. (See Section 3.2 for a discussion of the effects of fast charging.)

*Charge Time* — Obviously the charge time must be adequate for the charge rate selected including appropriate allowances for charging inefficiency. In general a charge time that results in return of 160 per cent of the standard capacity is the minimum acceptable. At nonstandard temperature conditions, this time may need to be lengthened. (Section 3.2 discusses selection of an adequate charging time.)

### 3.1.3.3.2 Rest Temperature and Time

Cell temperature, during a short rest period between charge and discharge, has a negligible effect on the value of actual cell capacity. However, elevating the rest temperature speeds up the process of self discharge which, if the rest period is lengthy, results in a reduction of the capacity retained by the cell and, hence, available for discharge. This subject is discussed in detail in Sections 3.1.3.4 and 3.3.

### 3.1.3.3.3 Discharge Conditions

There are two conditions from the discharge side that influence the actual capacity sufficiently to be considered in the derating process. These are:

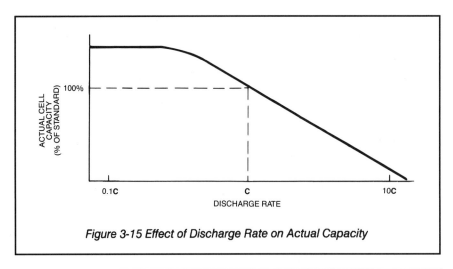

*Figure 3-15 Effect of Discharge Rate on Actual Capacity*

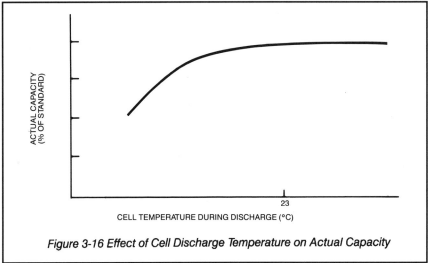

*Figure 3-16 Effect of Cell Discharge Temperature on Actual Capacity*

- Discharge Rate
- Discharge Temperature

*Discharge Rate* — Discharge rates affect actual cell capacity because of the increasing difficulties inherent in electrolyte mass transport and electrode reactions as the current density is increased. The influence of rate on actual cell capacity, at standard temperatures and at rates of discharge up to 0.25C, is insignificant. The relationship between discharge rate and cell capacity at discharge rates above this value is semilogarithmic with increasing rate as shown in Figure 3-15. The reduced capacity at increased discharge rates simply reflects a reduction in the amount of charge which is available at these increased rates. The energy represented by the capacity that is not dischargeable at those higher rates is not lost or dissipated. It remains in the cell where it may be discharged at lower rates.

*Cell Discharge Temperature* — Low cell temperature during discharge has an effect upon actual cell capacity. That relationship is shown in Figure 3-16.

### 3.1.3.3.4 Simultaneous Multiple Off-Standard Conditions

Actual cell capacity can still be estimated in those instances where several of the above charge and discharge conditions are off-standard at the same time. Since the

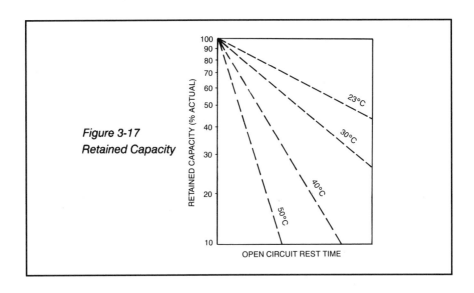

Figure 3-17
Retained Capacity

derating factors are reasonably independent of each other, multiplying them together gives a combined derating. For example, 90 per cent derating due to high discharge rate multiplied by 80 per cent derating due to high charge temperature yields an estimated actual cell capacity of 72 per cent of standard when both conditions are present.

Combining derating factors must be done with care because not all combinations of off-standard conditions are totally independent. Instead, some combinations work together to produce greater derating than predicted. This is especially true if the off-standard conditions are near the cell operating extremes. For example, assume that a derating for a cell temperature of –20°C at the standard 1C rate of discharge is 65 per cent, and the derating for a 5C discharge at the standard cell temperature of +23°C is 84 per cent. The combined effect of 5C and –20°C is not the calculated 55 per cent (65 per cent multiplied by 84 per cent), but a significantly lower 35 per cent of standard capacity. If the anticipated battery application involves combinations of conditions that are significantly off-standard, the application designer should request assistance from the manufacturer in estimating the derating from standard performance.

### 3.1.3.4  Retained Capacity

Capacity retention is a measure of the ability of a battery to retain stored energy during an extended open-circuit rest period. Retained capacity is a function of the length of the rest period, the cell temperature during the rest period, and the previous history of the cell. Capacity retention is also affected by the design of the cell. Nickel-cadmium cells are manufactured using a variety of electrode formulation processes. Each process is unique and electrodes made from each of these processes may possess different capacity retention characteristics. Other cell design parameters such as electrode spacing, electrode surface area, and separator material, also affect capacity retention.

*Self discharge* is the term used to describe the decay of retained capacity. The rate of self discharge of any particular cell design depends on the amount of retained capacity and the cell temperature.

Figure 3-17 illustrates the typical retained capacity of a nickel-cadmium cell, as a function of storage time and cell storage temperature, in terms of a per cent of the original actual capacity stored in the cell at the beginning of a rest period. Details of the self-discharge process and how it influences storage times are provided in Section 3.3.

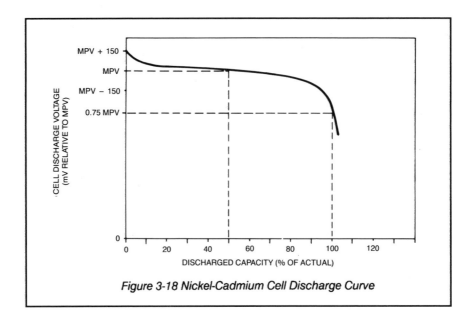

*Figure 3-18 Nickel-Cadmium Cell Discharge Curve*

## 3.1.4    SCALING THE DISCHARGE CURVE

The capacity and voltage discussions in Sections 3.1.2 and 3.1.3 can now be used to actually place voltage and capacity values on the curve of Figure 3-18. The redrawn curve, with voltage and capacity scales, appears as the Universal Discharge Curve shown in Figure 3-19.

The procedure for arriving at values for these scales for use on Figure 3-19 is described below. Worksheets for scaling the discharge curve to actual conditions are provided on some manufacturers' cell specification sheets.

APPLICATION PARAMETERS
Determine the operating conditions of interest for the specific application:

Cell Charge Temperature (°C)
Cell Discharge Temperature (°C)
Cell Discharge Rate (mA)

CAPACITY AND RUN TIME
1) Start with the Rated Capacity **C**. This is available from the specifications for the cell type as supplied by the manufacturer.
2) Determine the changes in capacity attributable to use of the cell at conditions different from the standard conditions.
a) Using the actual cell charge temperature with the derating curve from the Appendix or from the cell specification sheet, determine the derating factor (if any) for the difference in charge temperature.
b) Using the actual cell discharge temperature with the derating curve from the Appendix or from the cell specification sheet, determine the derating factor (if any) for the difference in discharge temperature.
c) Using the actual cell discharge rate with the derating curve from the Appendix or from the cell specification sheet, determine the derating factor (if any) for the difference in discharge rate.
3) Determine the actual capacity by multiplying the rated capacity by the product of the three derating factors calculated in Step 2. As discussed earlier, take care in interpreting the results if significant derating occurs in more than one parameter.

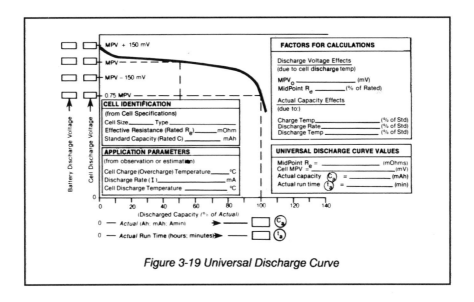

*Figure 3-19 Universal Discharge Curve*

4) Actual run time is the actual capacity determined in Step 3 divided by the discharge rate.

VOLTAGE

1) Obtain the value of $R_e$ at standard conditions from the cell specification sheet.
2) Using the cell discharge temperature and the appropriate curve in the Appendix or on the cell specification sheet, determine the effect of cell discharge temperature on cell resistance $R_e$.
3) Calculate the voltage drop in the cell due to internal resistance in mV as the product of the cell discharge rate in mA, $R_e$ at standard conditions in m$\Omega$, the derating factor for temperature effects on $R_e$, and 0.001 (to convert the units properly).
4) Read the midpoint open-circuit voltage ($MPV_o$) from the curve of midpoint voltage versus cell discharge temperature located in the Appendix or on the cell specification sheet.
5) The application midpoint voltage is the $MPV_o$ at the cell discharge temperature less the voltage drop calculated in Step 3.
6) The full capacity end-of-discharge voltage is 75% of the application midpoint voltage.

THE RESULT

Placing these values of actual capacity, discharge time, and cell voltage, in their proper position on the Universal Discharge Curve (Figure 3-19), produces a graph of the actual operating voltage versus discharged capacity and time. This graph will thus be a reasonable approximation of the discharge voltage curve to be expected from a relatively new battery at the set of operating conditions described. The next section will discuss the influence of various aging factors on battery performance.

## 3.1.5  APPLICATION PERFORMANCE AND OTHER OPERATING CHARACTERISTICS

The discharge performance of cells and batteries is of course dependent on their state of health in addition to the immediate environment or application factors considered in Sections 3.1.2 and 3.1.3. The performance that was scaled in Section 3.1.4 is

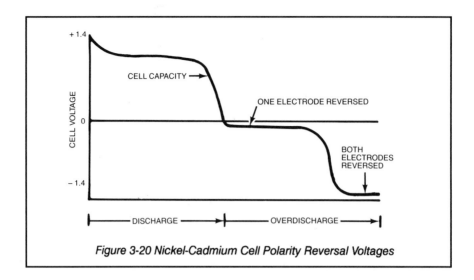

*Figure 3-20 Nickel-Cadmium Cell Polarity Reversal Voltages*

based on a cell in the as-new condition. As the cell is used its performance may change. A number of factors operate to change the characteristics of the cell over an extended period of time in use. These factors may all be grouped under the general heading of cell history—the electrical, thermal, and mechanical environments that the cell has experienced during its lifetime.

Some of the possible historical events which may degrade a cell's performance relative to Figure 3-19 are as follows:

- Repeated cell polarity reversal
- Excessive charge/overcharge rates (discussed in Section 3.2)
- High overcharge cell temperature
- Storage at elevated temperatures
- High discharge cutoff voltage
- Normal cell aging

The following discussions of the effects of cell history focus on its impact on discharge performance. The relationship between cell use history and cell life is explored in more detail in Section 3.4.

### 3.1.5.1   Repeated Cell Polarity Reversal

When a multicell battery is discharged completely, even small differences in the actual capacity of individual cells may cause one cell to reach complete discharge sooner than the rest. It is the actual capacity of this lowest capacity cell which thus determines the actual battery capacity.

Continuation of the discharge after the lowest capacity cell has reached zero volts will cause reversal of the terminal voltage of that cell because the cells of the battery are connected in series. As the cell polarity is reversed beyond a negative 0.2 volts it begins to generate gas internally. Generally, in a sealed nickel-cadmium cell the electrode that reverses first is the nickel (positive) electrode. The result is hydrogen generation. The sealed cell dissipates hydrogen very slowly and has a very limited amount of gas storage volume. Thus, frequent or extensive cell reversal will lead to an internal pressure sufficient to open the resealable safety vent. Application conditions causing repetitive or deep cell reversal may result in excessive venting which will eventually degrade the cell's performance.

The general shape of the cell voltage curve during discharge into polarity reversal is shown in Figure 3-20. The magnitude of the negative voltage appearing on the polarity reversed cell depends on the discharge rate as well as cell design and manufacturing process parameters.

Many nickel-cadmium cells are designed and manufactured to resist the adverse effects of cell reversal caused by the deep discharging of a multicell battery. In batteries consisting of a small number of cells that are fully charged between cycles, the cells are normally close enough in capacity to avoid significant permanent damage from cell reversals in deep discharges.

The important point relative to cell reversal is that the degradation is cumulative. The degree of degradation is dependent on both the depth (Ah) in reverse and the frequency of cell reversal. The depth of reversal is in turn dependent on the EODV and the actual capacity differences between the cells in the battery. The long-range detrimental effect on voltage and capacity performance increases as the EODV decreases or the spread in capacity increases.

The detrimental effects of reversal are the least when the rates of discharge during the reversal are the highest. This apparent paradox results because cells which experience terminal voltage reversal during high rate discharge still have significant amounts of charge remaining in the electrodes. (See Section 3.1.5.3.4 for cell polarity reversal voltage)

### 3.1.5.2  High Overcharge Cell Temperature and Voltage Depression

The effects of elevated charge temperature on the immediate cycle capacity of the cell have been discussed in Sections 3.1.2.1 and 3.1.3.3.1. Cells exposed to overcharge for extended periods of time, particularly at elevated cell temperatures, may also exhibit a phenomenon called *voltage depression*. This results in the cell voltage being depressed approximately 150 mV below the normally expected values calculated on Figure 3-19. This depression affects $E_o$ and is independent of discharge rate.

Voltage depression initially appears on the discharge voltage curve near the end of discharge. With extension of the overcharge time (non-discharge) of the cell, this depression progresses slowly toward the mid-point and beyond. Accompanying the depression in the voltage dimension of the curve is a slight increase in the capacity dimension as illustrated in Figure 3-21. Voltage depression is an electrically reversible condition and disappears when the cell is completely discharged and recharged (sometimes called *conditioning*). It thus appears only on the first discharge following a very extended overcharge. It will reappear if the extended overcharge is repeated.

The cause of voltage depression is continuous overcharging of the active material of the electrode. The effect is erased by discharging and recharging that portion of the active material which has experienced the extensive overcharge. For this reason the depressed voltage effect in the discharge curve is erased by the very act of observing it, when the discharge is carried beyond the second knee of the depressed curve. Complete discharge, and subsequent full charge, essentially restores the curve to its normal form.

The reversibility of voltage depression is probably the very characteristic that gives rise to the misnomer *memory*. When cells are subjected to continuous charge/overcharge with only modest discharges (repetitive or otherwise), the reversibility of the effect actually prevents voltage depression from occurring in that portion of the electrode active material which is cycled. The voltage depression phenomenon is, however, not erased from that portion of the electrode material which has been subjected to continuous overcharge but NOT discharged. In this situation, whenever the cell is discharged deeper than recent previous discharges and reaches the beginning of the previously uncycled material, the voltage may decrease 150 mV per cell. This may mislead the observer into believing that the discharge is at the knee of the normal discharge curve and erroneously concluding that the cell remembers and, thus,

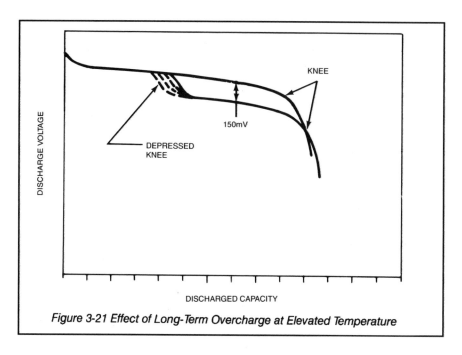

*Figure 3-21 Effect of Long-Term Overcharge at Elevated Temperature*

delivers only the amount of capacity previously used repetitively. Instead, the phenomenon is actually related only to extended overcharging and incomplete discharging, not repetitive shallow cycling. This is because that portion of the electrode material which has experienced overcharge and not been discharged for an extended period of time slowly shifts to a more inaccessible form.

The depressed voltage effect can, of course, cause loss of useful capacity in those applications where a high cutoff voltage prevents complete discharge of the minimum capacity cell in the battery. If voltage depression has occurred, complete discharge requires continuation down through the depressed knee to the voltage level which keeps all the electrode material active. An end-of-discharge voltage (EODV) on the minimum cell of, for example, 75 per cent of its MPV or less will accomplish that restoration with each discharge. This is further discussed in Sections 3.1.5.4 and 3.5.3.4.

### 3.1.5.3    Discharge Termination Voltage vs. Capacity

In actual use, the depth of discharge ordinarily experienced by the battery is not determined by voltage at all but by the use pattern of the product. For the majority of applications, all the available capacity is seldom used at any one time. The discharge is typically terminated in these products when the immediate task at hand is completed — for example a 5-minute shave with a shaver battery capable of 40 minutes, or a power outage requiring 48 seconds of emergency lighting from a system with a design capacity of 90 minutes. These discharge durations should be considered to be random time-controlled discharges and are not a function of battery voltage.

#### 3.1.5.3.1 *Available Cell Capacity as a Function of EODV*

Figure 3-4 indicates that less capacity is available if the end-of-discharge voltage (EODV) is increased, and more capacity will be available if the EODV is decreased. A close-up view of the knee of the discharge curve can demonstrate this effect on the available capacity of the cell, relative to the actual capacity.

Figure 3-22 illustrates this close-up view of the two extremes in the shape of the knee. This figure shows the amount of capacity available to various EODV values. Curve A in Figure 3-22 has a relatively sharp normal knee which results in only

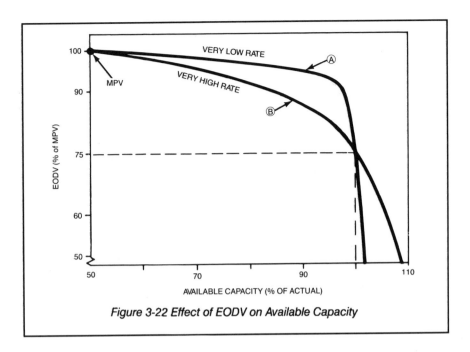

*Figure 3-22 Effect of EODV on Available Capacity*

a minor variation in available capacity with EODV's equal to or lower than 90 per cent of MPV. Curve B in Figure 3-22 represents a different curve shape resulting from some combination of higher discharge rate, extended overcharge, higher cell temperature, history, etc. Curve B illustrates the more significant variation of available discharge capacity which may be present in cells with a more rounded knee in their discharge voltage characteristic.

### 3.1.5.3.2 *Discharge Cutoff Voltage*

Battery-operated products often use more than one level or means of battery discharge load switching. First there is the primary switching by which the battery load is normally initiated and terminated. This load switching may be either manual (by a simple on/off or by a momentary type switch) or automatic (by electronic circuitry).

In addition some products contain a back-up circuit which may preempt the primary load switch by terminating the discharge at a predetermined battery voltage. These preemptive discharge termination techniques, which tend to determine the minimum EODV value (i.e., the maximum amount of discharge) in various applications, may be classified as either hard, semi-hard or soft.

Products having hard cutoffs contain circuitry that senses the battery voltage actually being delivered to the product and automatically terminates the battery discharge by electrically switching the battery load off when battery voltage falls below a predetermined value. A hard cutoff voltage is frequently designed into those products in which a low supply voltage from the battery would result in unsatisfactory load performance; the negative consequences of which might be undetectable to the user at the time. Some load circuits, particularly electronic loads, may possess such a nonlinear relationship between battery voltage and load current that they inherently display nearly hard cutoff characteristics, even without an actual battery-switch-out mechanism. Typical products which may use hard cutoffs are video cameras, transceivers and cordless telephones.

There is a second similar design technique used only in manually operated products. These products might be considered to have a semi-hard cutoff voltage. They are characterized by battery voltage sensing circuits similar to the hard cutoff case, but these circuits only signal the user that the battery is discharged. Further discharge is not prevented if the operator, for any reason, does not open the primary load switch. (See Section 3.5.3.3 on low state-of-charge sensing.)

The third category of applications inherently has a soft cutoff voltage. The soft cutoff case is illustrated by an application in which the user is the only judge of the amount of reduction of function, resulting from the decreasing battery voltage, which will be tolerated before deciding to terminate the discharge and recharge the battery. This is the cutoff employed by all appliances in which no additional circuitry has been provided beyond the primary load switch. Typical products in the soft cutoff category are power tools with momentary type load switches, and flashlights with simple on/off primary load switching.

There are two principal criteria which should be used to select discharge cutoff voltages for any of the three types of discharge termination. These criteria are:

- Required Application Voltage (RAV)
- Cell Polarity Reversal Voltage (CPRV)

### 3.1.5.3.3 *Required Application Voltage*

Required Application Voltage (RAV) is the lowest battery voltage which will produce acceptable product function. Its value is dictated by the system design of the end product. This voltage is the lower limit of the operating voltage range between battery supply voltage and end product load voltage.

Products are normally specified to operate nominally at the average delivered battery voltage. The average operating voltage in the case of the battery is approximately equal to the midpoint voltage (MPV). It is therefore helpful to think of the RAV lower voltage limit as a tolerance on the minimum side of that average operating voltage (i.e., some fractional portion of the nominal operating voltage).

As a general rule the capacity utilization of a single-cell battery will be very close to 100 per cent of the total battery (cell) capacity when the end product is designed to function satisfactorily down to a single-cell battery voltage of 75 per cent of the average operating voltage (0.75 multiplied by MPV). This results in a required end-product functional voltage tolerance of +15 per cent and -25 per cent from the nominal MPV operating value. Since MPV may be conveniently calculated by the procedures illustrated in Section 3.1.4, all the effects of cell design, as well as application discharge rate and cell temperature, are thus inherently compensated for by using the 0.75 MPV cutoff rule.

In the case of multicell batteries, 100 per cent utilization of the total capacity capability is achieved with even less down-side voltage tolerance requirements. This is due to the effect of the sharp drop in battery voltage which occurs when the minimum capacity cell in the battery reaches the end of its capacity and initiates the drop to EODV.

The relationship between battery voltage and product performance may be adjusted during the design of the end product by increasing or decreasing the number of cells in the battery (battery supply voltage). It may also be adjusted by increasing or decreasing the response of the end product at specific battery voltages. The battery supply voltage can also be increased in the case of high-rate, low-run-time applications by the use of a low $R_e$ design cell, or a larger cell with higher capacity which thus inherently has a lower $R_e$.

### 3.1.5.3.4 *Cell Polarity Reversal Voltage*

Although it may be convenient to think of multicell battery voltage in terms of mean voltage per cell (battery voltage/number of cells), the mathematics of the cell polarity reversal phenomenon illustrated in Section 3.1.5.1 result in a fundamental difference from this convenient and simple relationship. The voltage delivered by the nickel-cadmium cell, following the knee of the curve, normally falls off with extreme rapidity. Because of this characteristic, small differences in the actual capacity of the various cells in the battery may result in a low and possibly reversed voltage for the cell with the minimum capacity before the remaining cells in the battery have even reached the knee of their particular voltage curve. The limiting cell-reversal scenario might thus be comprised of all cells save one at or above their knee while the lowest capacity cell is at zero or even a slight negative voltage.

Polarity reversal of the minimum capacity cell can be prevented, if necessary, by limiting the multicell battery EODV to approximately:

$$CPRV_1 = (MPV - 100)(n - 1) \text{ millivolts}$$

Where:  CPRV = minimum battery voltage which avoids cell reversal
MPV = midpoint voltage for the individual cell
n = number of cells in battery

This equation describes the battery voltage with one cell at zero and the other $(n - 1)$ cells at a point just prior to their voltage knee which is generally located at least 100 mV below midpoint. The result is a conservative estimate of the discharge termination battery voltage that prevents driving any cell in the battery into reverse polarity. As an example, consider a 5-cell battery which has a nominal voltage of 6 volts. The MPV for the individual cells is 1.2 volts and $CPRV_1$ would be $(1200 - 100)(5 - 1)$ or 4.4 volts.

A more generous determination of CPRV allows a greater average decrement of voltage below midpoint for the $(n - 1)$ cells, for example 150 mV each. It also recognizes that the low cell must reverse to at least 200 mV prior to any detrimental effects. Taking both of these allowances into account, a more liberal reversal prevention voltage is:

$$CPRV_2 = [(MPV - 150)(n - 1) - 200] \text{ millivolts}$$

For the example cited above, the result would be $[(1200 - 150)(5 - 1) - 200]$ or 4.0 volts.

Because the damaging effects of cell reversal depend on the cumulative amount (amp-hour magnitudes) of capacity in reverse and each reversal may utilize only a fraction of the total tolerance for any one cell, the minimum operating voltage may be further reduced by:

a) Use of cell designs with greater tolerance for reversal.
b) Recognition that use of the end-product may infrequently rely on cutoff voltage as the actual discharge termination.
c) Use of cells with tightly grouped capacity distributions for making each battery.
d) Accepting a reduction in battery life attributable to cumulative cell degradation due to deep reversals.
e) Considering the general rule that higher rate discharges have softer knees with greater decrements below their MPV, further reducing the value of CPRV and thus minimizing the possibility of cell reversal.

Only the end-product designer is in a position to balance the values of RAV and CPRV for the various requirements of cutoff voltage. It is helpful, however, to understand that the MPV is the key voltage from which all these factors may be considered. MPV is the average voltage delivered under the actual load and therefore determines the average functional performance of the end product in addition to its relationship to minimum acceptable product performance. MPV also determines the initial discharge voltage delivered to the end product, approximately $[n \times (MPV + 150)]$ mV. MPV also fixes the maximum battery voltage at which any one cell could possibly be placed in reverse polarity.

The low voltage tolerance of the load need not be designed any lower than either 0.75 multiplied by MPV or the selected value of CPRV, whichever is higher, in order to utilize the maximum useful capacity of the battery.

## 3.1.6    CELL DESIGN FACTORS

The design of the cell can impact both the discharge voltage and the capacity available for a given load under specific operating conditions. The principal elements of the cell that affect discharge performance include electrode type and dimensions, current collectors, separator, and electrolyte. Variations in these design elements are used by many manufacturers to produce different cell types with performance characteristics that match a variety of applications.

### 3.1.6.1    Electrode Dimensions

The thickness and surface area of an electrode have a significant effect on the cell's ability to deliver voltage at high discharge rates and to produce maximum capacity. Normally, to maximize the capacity of a cell, the electrode is made rather thick and thus correspondingly shorter in coiled length. This thick electrode provides a cell with more active material than does a thin electrode of the same total volume, and, because of that additional active material, the cell will deliver more capacity. However, the thick electrode has a high $R_e$ and cannot deliver voltage at high discharge rates as well as the thin electrode of the same total volume and chemistry. For a specific current level, the surface current density in the thick electrode, with its smaller surface area, is higher than that of the thin electrode which has the larger surface area. Accordingly, for high capacity requirements the electrodes are designed thicker but with less surface area. For high-rate discharge service the electrodes are designed to be thin and have a large surface area, even at the expense of additional active material.

The thickness and surface area of an electrode also have an effect upon its ability to accommodate high overcharge rates. The oxygen evolved in overcharge at the positive electrode must be reduced and recombined at the negative electrode. An electrode with a large surface area will reduce the generated oxygen more readily and at a lower pressure than an electrode with a small surface area. For this reason, fast-charge cells are normally designed with thin electrodes having large area.

### 3.1.6.2    Current Collection Means

To achieve good voltage regulation at very high discharge rates, all parts of the cell must be designed for low internal resistance. One way to achieve low $R_e$ is to provide multiple parallel paths for current to flow from the electrode to the cell terminals. This is accomplished by using a current collector that contacts the electrode spiral in many places.

### 3.1.6.3    Separator

The separator fulfills a number of functions in the operation of a cell. These include:

- Maintaining electrical separation of the electrodes.
- Wicking the electrolyte to and from the electrodes while immobilizing that portion of the electrolyte not in the electrodes.
- Providing low resistance conduction of OH⁻ ions from one electrode to the other during charging and discharging of the cell.
- Providing permeability for oxygen gas to migrate rapidly and easily from the positive to the negative electrode during overcharge.
- Preventing or retarding the migration of cadmium species from the negative to the positive electrode.

In performing these functions a number of separator design trade-offs affect the performance of the cell.

A thicker and less porous separator improves the life of the cell. A thin separator reduces the $R_e$ and increases the volume available for plate active materials as well as improving the oxygen gas transfer between electrodes, thus enhancing overcharge rate capacity.

Cells may be designed for extended high-temperature operation by using a polypropylene separator along with material changes in other life-limiting components. For normal cell temperatures a polyamide material is generally specified. The polyamide material has better electrolyte wettability and therefore wicks the electrolyte better than polypropylene. It thus provides higher ion conductivity, resulting in lower internal resistance and better capacity stability over the life of the cell.

### 3.1.6.4   Electrolyte

The choice of electrolyte concentration and additive selection impacts cell life and performance. Most nickel-cadmium cells are designed with a potassium hydroxide electrolyte having a concentration ranging from 29 per cent to 37 per cent. The concentration used depends upon the application temperature of the battery and the trade-off between performance and life. A variety of additives may also be used in the electrolyte to enhance performance in particular types of applications.

### 3.1.6.5   Normalizing $R_e$

The effects of cell design, aspect ratio of the cell shape, plate manufacturing methods, etc., on the internal resistance $R_e$ may be seen in a sharper perspective by normalizing the $R_e$ value relative to cell capacity. To accomplish this, the resistance is expressed in terms of the voltage decrease for each **C** rate of increase of discharge rate (mV/C amperes). Thus an $R_e$ of 30 milliohms for a 0.5 Ah cell becomes 15 mV/C rate (30 $\times$ 0.5). For example, a discharge current of 8C rate would result in a discharge voltage which was 120 mV/cell below the effective open circuit voltage $E_0$. Another cell design with an $R_e$ of only 7 milliohms in a 4 Ahr cell becomes 28 mV/C rate (7 $\times$ 4). This 7 milliohm cell is thus almost twice as high in effective resistance as the 30 milliohm cell when the effect of its larger capacity is normalized out of the measurement. This method of expressing $R_e$ better illustrates the effects of cell design and is frequently convenient to use because it is a more constant value which indicates overall cell technology and relative power delivery capability.

## 3.1.7 SUMMARY

The discharge characteristics which describe sealed nickel-cadmium cell performance are capacity and voltage. Capacity, measured in terms of ampere-hours (or milliampere-hours), is dependent upon cell size, the cell design and construction, the effectiveness of the charge, the cell temperature of the application, the open circuit time between charge and discharge, the discharge rate, the end-of-discharge voltage, and the previous history of the cell in cycling, overcharging, or idleness.

The average discharge voltage is dependent upon the cell size and construction, the discharge rate, the cell temperature of the application, and the previous history of the cell in cycling or idleness.

These two characteristics, capacity and voltage, were discussed in detail in Sections 3.1.2 and 3.1.3. Techniques for determining and illustrating their approximate new cell values in any application situation were described in Section 3.1.4. The effects of battery usage on discharge parameters were indicated in Section 3.1.5. The impact of cell design on performance was briefly considered in Section 3.1.6.

# 3.2 Charging

After assessing whether sealed nickel-cadmium batteries will supply an application's electrical requirements, the next major decision facing a product designer is selection of the approach to be used in charging the battery. Proper charging is a key to success with any battery application. This section explains the basic approaches to charging sealed nickel-cadmium batteries and provides suggestions on matching chargers to applications. Since proper charger design or selection can be critical to battery performance, application designers, after reviewing the information in this section, may still want to work closely with both cell and charger manufacturers to obtain the optimum charging system for the application.

## 3.2.1 INTRODUCTION

Nickel-cadmium batteries are charged by applying a current of proper polarity to the terminals of the battery. The charging current can be pure direct current (DC) or it may contain a significant ripple component such as half-wave or full-wave rectified current.

This section on charging sealed nickel-cadmium batteries refers to charging rates as multiples (or fractions) of the $C$ rate. The $C$ rate was defined in Section 2.2.5 as the rate in amperes or milliamperes numerically equal to the capacity rating of the cell given in ampere-hours or milliampere-hours. For example, a cell with a 1.2 ampere-hour capacity has a $C$ rate of 1.2 amperes. The $C$ concept simplifies the discussion of charging for a broad range of cell sizes since the cells' responses to charging are similar if the $C$ rate is the same. Normally a 4 Ah cell will respond to a 0.4 amp ($0.1C$) charge rate in the same manner that a 1.4 Ah cell will respond to a 0.14 amp (also $0.1C$) charge rate. These $C$ rate charging currents can also be categorized into descriptive terms, such as standard-charge, quick-charge, fast-charge, or trickle-charge shown in Table 3-3.

| METHOD OF CHARGING | CHARGE RATE | | RECHARGE TIME* (HOURS) | CHARGE CONTROL |
|---|---|---|---|---|
| | MULTIPLES OF C-RATE | FRACTION C-RATE | | |
| STANDARD | 0.05C 0.1C | C/20 C/10 | 36-48 16-20 | NOT REQUIRED |
| QUICK | 0.2C 0.25C 0.33C | C/5 C/4 C/3 | 7-9 5-7 4-5 | NOT REQUIRED |
| FAST | C 2C 4C | C 2C 4C | 1.2 0.6 0.3 | REQUIRED |
| TRICKLE | 0.02-0.1C | C/50-C/10 | Used for maintaining charge of a fully charged battery. | |

*RECHARGE TIME = STANDARD TIME TO FULLY CHARGE A COMPLETELY DISCHARGED BATTERY AT 23°C

*Table 3-3 Definition of Rates for Charging*

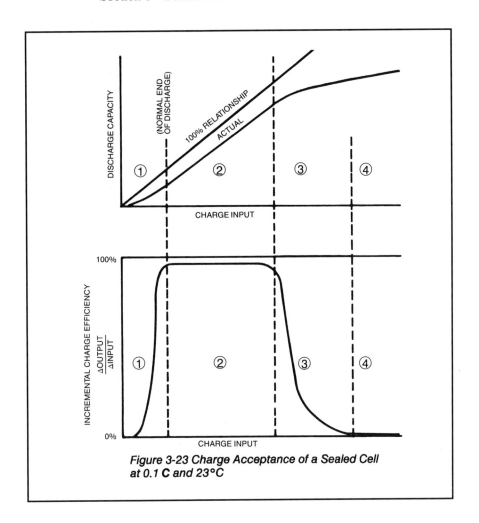

*Figure 3-23 Charge Acceptance of a Sealed Cell at 0.1 C and 23°C*

## 3.2.2 CHARGING EFFICIENCY

When a nickel-cadmium battery is charging, not all of the energy input is converting the active material to a usable (dischargeable) form. Charge energy also goes to converting active material into a unusable form, generating gas, or is lost in parasitic side reactions. The term *charge acceptance,* which characterizes charging efficiency, is the ratio of the dischargeable capacity obtained to the charge input.

The top curve of Figure 3-23 shows the dischargeable capacity (charge output) as a function of the charge input for a sealed cell starting from a completely discharged state. The ideal cell, with no charge acceptance losses, would be 100 per cent efficient: all the charge delivered to the cell could be retrieved on discharge. But nickel-cadmium cells typically accept charge at different levels of efficiency depending upon the state of charge of the cell, as shown by the bottom curve of Figure 3-23. Four successive types of charging behavior — Zones 1, 2, 3 and 4 in Figure 3-23 — describe this performance. Each zone reflects a distinct set of chemical mechanisms responsible for loss of charge input energy.

In Zone 1 a significant portion of the charge input converts some of the active material mass into a non-usable form, i.e. into charged material which is not readily accessible during medium or high-rate discharges, particularly in the first few cycles. In Zone 2, the charging efficiency is only slightly less than 100 per cent; small

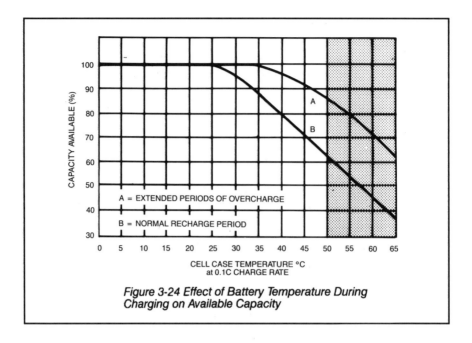

*Figure 3-24 Effect of Battery Temperature During Charging on Available Capacity*

amounts of internal gassing and parasitic side reactions are all that prevent the charge from being totally efficient. Zone 3 is a transition region. As the cell approaches full charge, the current input shifts from charging positive active material to generating oxygen gas. In the overcharge region, Zone 4, all of the current coming in to the cell goes to generating gas. In this zone the charging efficiency is practically zero. Overcharge is further discussed in Section 3.2.4.

The boundaries between Zones 2, 3, and 4 are indistinct and quite variable depending upon cell temperature, cell construction, and charge rate. The level of charge acceptance in Zones 1, 2, and 3 is also influenced by cell temperature and charge rate.

### 3.2.2.1   Effect of Temperature

Both the charge acceptance and the actual capacity of a cell charged at elevated temperatures are lower than for a cell charged at room temperature. Even though the total charge input to a hot cell may exceed its rated capacity by many times, the cell accepts only a portion of the charge that it would accept at room temperature.

As a nickel-cadmium sealed cell approaches its maximum state of charge (Zone 3), an increasing portion of the charging current goes to generating oxygen gas inside the cell and less goes to raising the cell's state of charge. At higher cell temperatures gas generation begins at lower states of charge. This reduces actual capacity at elevated charge temperatures even though the cell is in overcharge (Zone 4). This is illustrated in Figure 3-24, 3-25, and 3-26. The effect of this phenomenon on dischargeable capacity was discussed in Section 3.1.3.3.

Figure 3-24 shows the capacity reduction that occurs as the cell temperature increases. The effect is most pronounced when the charge time is not adjusted to account for the higher temperatures. However, even with extended periods in overcharge, the cell still suffers very substantial losses in capacity at elevated temperatures.

Figure 3-25 indicates the effect of cell temperature on incremental efficiency and dischargeable capacity while maintaining the charge at a fixed 0.1C rate. The charge acceptance under standard conditions (23°C) is used as a reference. As the charging temperature increases to either 45°C or 60°C, both the charging efficiency

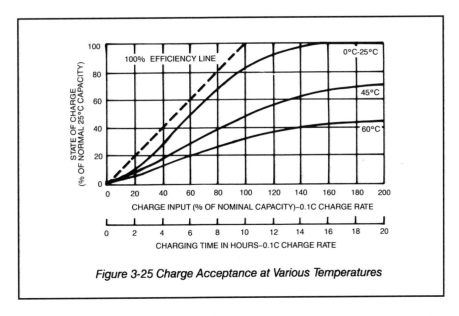

Figure 3-25 *Charge Acceptance at Various Temperatures*

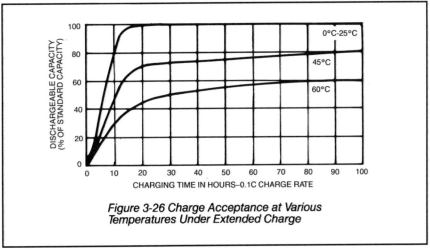

Figure 3-26 *Charge Acceptance at Various Temperatures Under Extended Charge*

and actual cell capacity decrease. With a cell temperature of 45°C and a charge input of 200 per cent of standard capacity, the actual capacity is less than 70 per cent of room-temperature capacity. Similarly, with a cell temperature of 60°C and a charge input of 200 per cent of standard capacity, the actual capacity barely reaches 45 per cent of the room-temperature value. However, a cell charged at temperatures between 0°C to 25°C accepts charge effectively and yields 100 per cent of standard capacity with a charge input of 160 per cent of its standard capacity.

The effect of extended charging times at a constant 0.1C charge rate is shown in Figure 3-26 which continues the charge input time scale of Figure 3-25 to 100 hours. The cell charged at 23°C, being charged to 100 per cent of standard capacity within 20 hours, does not increase its state of charge by charging beyond 20 hours. However, the 45°C cell, charged to only about 70 per cent of standard capacity in about 20 hours, gains capacity very gradually by additional charging. Similarly, the cell charged at 60°C also increases its dischargeable capacity with increased charge input. The charge input required to reach overcharge for a 60°C cell may be as much as one thousand per cent of standard capacity. Even then the cell never reaches its

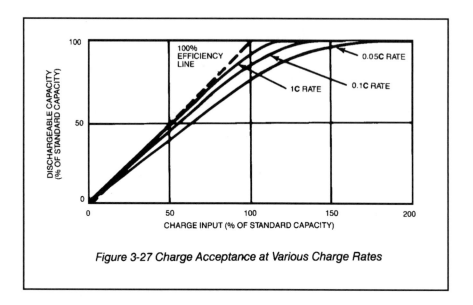

*Figure 3-27 Charge Acceptance at Various Charge Rates*

room-temperature capacity. As the active materials in a cell are converted to the charged state, the dischargeable capacity ultimately reaches the limit defined by the actual cell capacity of curve A in Figure 3-24.

### 3.2.2.2 Effect of Charge Rate

Figure 3-27 shows the effect of charge rate on dischargeable capacity for various charge inputs. Charge rates much below 0.05C do not enable the cell to reach full capacity. Charge rates higher than 0.1C enhance charge efficiency for cells that are capable of charging at higher than standard-charge rate, i.e. quick and fast-charge cells. For example a fast-charge cell charged at the 1C charge rate approaches standard capacity at an input of about 120 per cent. However, as discussed in Section 3.2.6, the fast-charge cell cannot sustain this high rate in overcharge.

### 3.2.2.3 Effect of Cell Construction

Cell design and construction can affect charge efficiency at elevated temperatures. Modifying the chemical composition of the electrodes and the electrolyte can improve the charge characteristics of cells specified for high-temperature use. This is not done to all cells as it may also degrade low-temperature cell capacity and fast-charge capability, and raise internal resistance. The cell specification data shows actual cell capacity as a function of charge temperature for each cell type.

### 3.2.3 CELL PRESSURE, TEMPERATURE AND VOLTAGE INTERRELATIONSHIPS

The relationship among cell pressure, temperature, and voltage during charge and overcharge is especially important, particularly at faster charge rates. Figure 3-28 represents, on one graph, the relationship between charge input and the temperature, pressure, and voltage of a typical cell, using a fixed charge rate of 0.1C at a 23°C ambient temperature.

Many sealed nickel-cadmium cells can be charged continuously at their cell specification overcharge rate by a simple constant-current charger without consuming or releasing any of the materials in the cell. The cells are designed so that, at the specification overcharge rate, the net electrochemical reaction occurring at one elec-

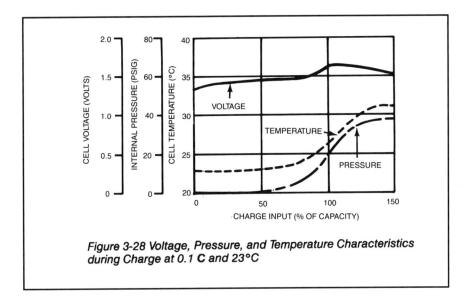

*Figure 3-28 Voltage, Pressure, and Temperature Characteristics during Charge at 0.1 C and 23°C*

trode is exactly opposite to that occurring at the other electrode, as was discussed in Section 2.3.2.1.

The cell pressure is driven by the quantity of gas in the cell. The pressure stays very low during most of the charge process, but starts to rise as the cell approaches full charge. Initially the bulk of the charge input goes into charging the active materials in the electrodes, and very little goes into gas-generating reactions. As the positive electrode approaches full charge, more and more charge energy goes to generating oxygen increasing the pressure in the cell. When the cell is fully charged, the pressure reaches equilibrium and ceases to rise. This is the stabilized overcharge condition where the rate of oxygen liberation at the positive balances the recombination rate at the negative electrode.

The cell temperature tracks the cell pressure but lags it slightly. During the major portion of the charge time, it increases only slightly. As the cell approaches full charge and oxygen generation increases at the positive electrode, the pressure increases and recombination of oxygen increases at the negative electrode. Since the oxygen recombination reaction is exothermic, increased recombination generates heat and the cell temperature rises. Eventually the cell reaches a stabilized pressure and temperature which are dependent on the heat transfer characteristics of the cell, the battery configuration, its container, the ambient temperature, and the charge rate.

Cell voltage also remains rather constant throughout most of the charging. It rises slightly as the cell approaches full charge but drops again in overcharge.

The effect of charge rate on the three parameters of cell pressure, cell temperature, and cell voltage is demonstrated dramatically by comparing Figures 3-28 and 3-29. Figure 3-28 shows the profiles of the same parameters for a 0.1C charge rate as Figure 3-29 shows for a 1C charge rate (ten times faster). Understanding the relationships between cell pressure, cell temperature, and cell voltage establishes some of the fundamental considerations in making nickel-cadmium cell and charger application decisions.

## 3.2.4   OVERCHARGE

*Overcharge* is the normal continued application of charging current to a battery after the battery has reached its maximum state of charge. It impacts the steady-state

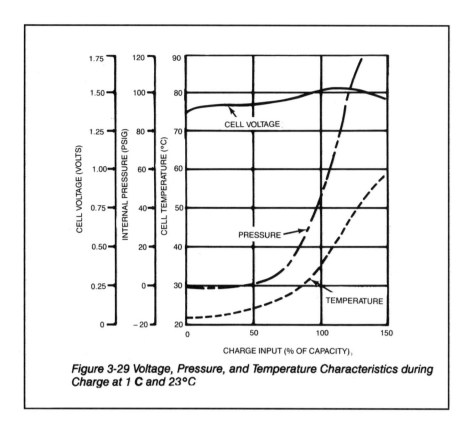

*Figure 3-29 Voltage, Pressure, and Temperature Characteristics during Charge at 1 C and 23°C*

values of pressure, temperature, and voltage. As discussed earlier, continued charging of fully charged cells causes the oxygen pressure to stabilize at an elevated level within the cells. The magnitude of the pressure increase depends primarily on the overcharge rate. Along with this rise of pressure comes an increase in cell temperature. Pressure, temperature, and voltage ultimately reach equilibrium in overcharge. Since cells are designed to reliably handle continuous overcharge at their cell specification rate, it is not an adverse condition. Overcharge is simply a term commonly used to describe the normal continuation of charge after the cell is fully charged.

Standard-charge cells may overcharge at rates up to 0.1C. Quick-charge cells, designed to withstand higher overcharge rates for an extended time, normally charge at rates up to 0.33C. Fast-charge cells, those that may charge at 1C to 4C rates, require special charger systems that automatically end high-rate charging. No matter which type of cell is involved, overcharge at rates above the cell specification rate may result in excessive temperatures and venting, and is therefore abusive to sealed nickel-cadmium cells.

The electrode design of fast-charge and quick-charge cells speeds recombination of oxygen at the negative electrode decreasing the pressure in the cell at any given rate. However, fast-charge rates still generate oxygen too rapidly to continue into overcharge without venting. Therefore, the charge rate must be reduced when the cell approaches full charge.

### 3.2.4.1   Tafel Curves

When a sealed nickel-cadmium cell is in overcharge within the cell specification rate, the internal pressure, temperature, and voltage of the cell eventually reach equilibrium. At equilibrium, the voltage/current/temperature relationship is quite predictable

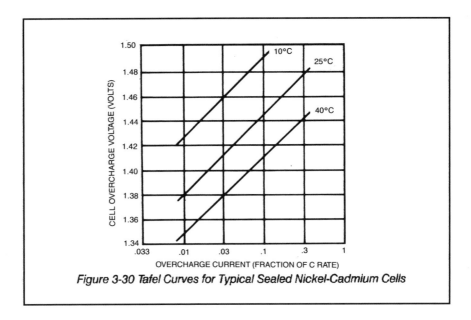

*Figure 3-30 Tafel Curves for Typical Sealed Nickel-Cadmium Cells*

from cell to cell. The characteristic of cell voltage vs. overcharge current, called the *Tafel Curve*, is a function of cell temperature as shown in Figure 3-30. Some of the charger circuits described later in this section use this characteristic as one factor in determining temperature compensation of their voltage sensing circuit.

### 3.2.4.2 Battery Overcharge Temperature

After a sealed nickel-cadmium battery has reached overcharge, all electrical input is converted to heat within the cells. As with any other body that transfers internally generated heat to its surroundings, the temperature of a battery in overcharge rises above the ambient temperature by an increment determined by the power input. Ultimately, the cells reach thermal equilibrium at a temperature that balances heat transfer to the environment with the heat generated within the cell through oxygen recombination. The overcharge temperature of a sealed nickel-cadmium battery is a major factor in determining battery life, as discussed in Section 3.4. Therefore, it is important to consider the factors determining battery overcharge temperature.

The temperature of the nickel-cadmium battery in overcharge is influenced by:

- Ambient temperature
- Overcharge current magnitude
- The heat transfer coefficient of the battery

These factors are discussed below as they relate to battery overcharge temperature.

The most efficient method of removing heat from a battery is through thermal conduction. Ideally, batteries in overcharge would use thermally conductive bonding of each cell to a heat sink maintained at a moderate temperature. This approach is seldom acceptable from the standpoint of both cost and space. Consequently, removing overcharge heat from a battery most often depends upon two less-efficient forms of heat transfer: convection and radiation.

For these modes of heat transfer, critical factors are the temperatures of the air and of any objects in the vicinity of the battery. In some applications, the battery is mounted inside equipment with poor ventilation or with other heat sources present.

The air inside such equipment and the walls of the compartment may be heated by the battery or other sources. This can cause the battery temperature to rise significantly more than if the battery were located in the room where the equipment is located. In estimating battery overcharge temperature, the immediate battery ambient temperature is the starting point to which is added the temperature rise caused by overcharge heat generation.

For convective heat transfer, the relationship between battery temperature rise and overcharge heat generation is linear. Though radiative heat transfer is nonlinear, it is, in reality, surprisingly close to linear over the relatively small range of battery temperature rise encountered in most applications. Consequently, for the combination of convection and radiation, the overcharge temperature rise ($\Delta T$) of a sealed nickel-cadmium battery is essentially proportional to the heat that must be dissipated. Since all of the energy input to a battery in overcharge is converted to heat, the temperature rise is proportional to the power input. This relationship is expressed below:

$$\Delta T \, (°C) = R_T \times P$$

Where:   $R_T$ = thermal resistance, battery to ambient (°C/watt)

   $P$ = overcharge power input (watts)

The overcharge power input is the product of battery overcharge voltage and overcharge current ($P = E_{oc} \times I_{oc}$). The overcharge voltage of the battery, within cell specification overcharge limits, is typically 1.45 volts per cell. The overcharge power input to a battery is thus:

$$P(\text{watts}) = E_{oc} \times I_{oc}$$
$$= n \times 1.45 \times I_{oc}$$

Where:   $E_{oc}$ = battery overcharge voltage

   $I_{oc}$ = overcharge current (amps)

   $n$ = number of cells in battery

   1.45 = typical cell overcharge voltage (volts)

Figure 3-31 allows estimation of the thermal resistance ($R_T$) associated with combined convective and radiative heat transfer to the surrounding ambient from a battery in a rectangular case. Entering the plot with the total surface area of the battery case (calculated in either square centimeters or square inches) returns the corresponding value of $R_T$ on the Y axis. The $R_T$ obtained from this curve will be reasonably accurate for typical battery case materials such as metals and solid plastics of practical thickness. The thermal resistance from the inner surface to the outer surface of a typical (nonfoamed) plastic case wall is relatively small. The two major elements of thermal resistance encountered in dissipating battery-generated heat are the thermal resistances from the cells to the case and from the case to the ambient. Of these two, the resistance from the case to ambient given by Figure 3-31 normally dominates.

With values for the overcharge power input and thermal resistance known, the battery temperature rise becomes:

$$\Delta T(°C) = R_T \times P$$

This temperature rise added to the battery ambient temperature is the predicted temperature at which the battery will operate in overcharge. Because of its simplicity, this heat transfer model often fails to account for some features specific to the application. Therefore, the calculated battery temperature for an encased battery should be considered a preliminary value. It should be confirmed by an actual temperature measurement made on a system prototype.

Several techniques, e.g. increasing battery surface area or eliminating the battery case, can be used to reduce overcharge temperature rise. Without the constraints

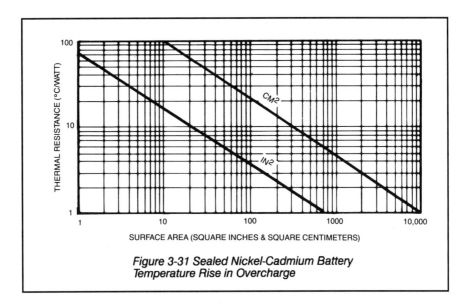

**Figure 3-31 Sealed Nickel-Cadmium Battery Temperature Rise in Overcharge**

of a case, better heat transfer may be attained by physically separating each cell. This way, air may flow freely over all cells, and more of the surface of each cell can transfer heat away. If a battery must be encased, in most situations, the minimum overcharge temperature occurs with the cells connected end to end. At the other extreme, the greatest temperature rise results when closely packed cells are foamed into a case approximating a cube. The foam impedes heat transfer from the cells to the case and the cubic battery shape minimizes the surface area for transferring heat from the case to its surroundings. A feasible design for a given application will probably lie somewhere between the extremes cited above. Nevertheless, design efforts that minimize the battery overcharge temperature will be rewarded by increased battery life.

Several general observations regarding overcharge temperature rise that are important in designing a battery into an application are summarized below. Each of these observations is implicitly accounted for in the previous equation.

- The battery overcharge temperature rise is essentially proportional to overcharge current ($\Delta T = R_T \times E_{oc} \times I_{oc}$). Battery temperature remains near ambient temperature, relatively independent of charge rate, until the battery approaches full charge. The battery temperature then transitions to the overcharge value where, for example, the temperature rise of a quick-charge battery overcharged at the 0.3C rate is three times the temperature rise of the same battery overcharged at the standard 0.1C rate, as shown in Figure 3-32.
- The number of cells in a battery affects overcharge temperature rise. A battery consisting of ten cells of a given size has a greater temperature rise than a battery of five cells of the same size and in the same configuration when both are overcharged at the same rate. This occurs because the input power in overcharge is proportional to the number of cells in the battery, but the battery surface area increases at a rate which is less than proportional to the number of cells. In summary, the battery with the larger number of cells has a greater overcharge input power per unit of battery surface area; consequently, it will have a greater temperature rise.
- If the same relative charge rate, such as 0.1C, is maintained, an increase in cell capacity usually increases overcharge heat generation proportionately but there is a less than proportionate increase in surface area. As a result, battery temperature increases with battery capacity when the same relative charge rate, e.g. 0.1C, is maintained.

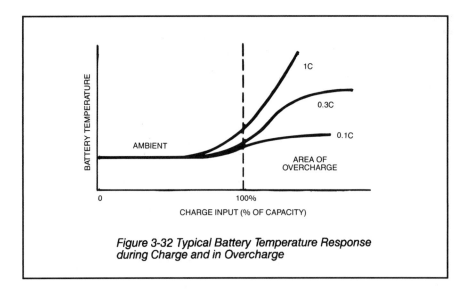

*Figure 3-32 Typical Battery Temperature Response during Charge and in Overcharge*

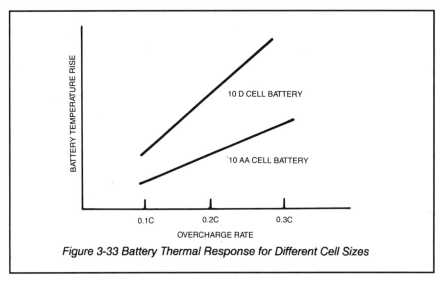

*Figure 3-33 Battery Thermal Response for Different Cell Sizes*

This is shown in Figure 3-33 where the overcharge temperature rise of a battery of ten D cells is greater than the overcharge temperature rise of a battery of ten AA cells.

### 3.2.4.3   Need for Charge Control

The cell internal pressure and temperature can rise considerably, particularly when the overcharge rate is high. In many cases the charge current is low enough that the cell is undamaged, even with years of uninterrupted overcharge. In other applications, the charge current is too high for continuous overcharge because either the cell pressure or temperature could rise high enough to damage the cell, limiting its life. For these applications, charge control is required. This is the subject of Section 3.2.6.

## 3.2.5   CHARGING WITHOUT FEEDBACK CONTROL: CHARACTERISTICS AND METHODS

In many applications for sealed nickel-cadmium cells and batteries, the charge current is low enough and the ambient temperature surrounding the battery is such that continuing the same charge rate into overcharge is quite acceptable. Where this is possible, the cost and complexity of feedback-controlled chargers is unnecessary.

The following sections discuss charging for applications where batteries may remain in continuous overcharge, even for years, at a constant charge rate.

### 3.2.5.1 Cell Capabilities

The most frequently used and most economical method for charging sealed nickel-cadmium batteries is the application of a continuous charge current that is within the cell specification overcharge capability. Typical sealed nickel-cadmium cells are available in two overcharge capabilities:

- Standard-charge or overnight charge capability
- Quick-charge capability (rates higher than the standard charge rate)

Fast-charge cells have the same overcharge capability as quick-charge cells. They can be charged more rapidly than either quick-charge or standard-charge cells, but this charge rate must be reduced in overcharge.

#### 3.2.5.1.1 Standard Cells

Standard-charge cells are designed to be continuously charged at rates up to 0.1C. Normally, a cell charged at the 0.1C rate is fully charged within 16 to 20 hours. In a few special instances 24 hours or more may be required for a full charge. Charge rates below 0.05C are normally discouraged, particularly in applications at elevated temperatures or where frequent discharges are required, as the cell may not achieve full performance.

#### 3.2.5.1.2 Quick-Charge and Fast-Charge Cells

Quick-charge and fast-charge cells permit continuous charging at rates up to 0.33C. This brings a discharged cell to full performance in as little as 4 to 5 hours. They can then continue indefinitely with an uninterrupted overcharge at that rate. The maximum overcharge rate for a quick-charge cell is the 0.33C rate. Obviously, quick-charge cells may be charged at lower rates, such as 0.2C, with the consequent lengthening of the time until the cells are charged. These lower rates are often used to reduce overcharge temperature rise in large battery packs.

Fast-charge cells may not be overcharged at the fast-charge rate. Their maximum overcharge rate is shown in Appendix A or in the pertinent cell specification sheet, but it is typically set at 0.33C or less.

### 3.2.5.2 Considerations for Charging Without Feedback Control

Charging without feedback control demands care to ensure that the charge rate and the cells are suited to the application's environmental requirements. If the ambient temperature is low, the cell internal pressure may rise too high in overcharge due to diminished oxygen recombination capability at the negative electrode. Figure 3-34 shows the impact of cell temperature upon stabilized internal cell pressure in overcharge. Therefore, if the battery will be exposed to low temperature environments charge control may be required. (Section 3.2.7.5 discusses low temperature charging.)

If the ambient temperature is high, the battery temperature in overcharge, (the combination of the overcharge temperature rise and the elevated ambient) may warrant attention, especially for large battery packs with large cells, as Section 3.2.4.2 showed. Continued exposure of cells to high temperature may shorten the battery's life. Therefore, either charge control to reduce overcharge temperature rise or installation of special cells designed for high-temperature applications may be necessary to obtain acceptable battery life.

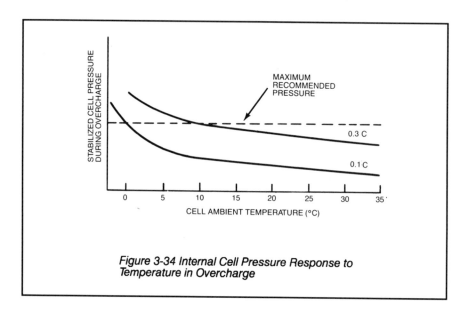

Figure 3-34 *Internal Cell Pressure Response to Temperature in Overcharge*

## 3.2.6   CHARGING WITH FEEDBACK CONTROL: CHARACTERISTICS AND METHODS

Although constant-current charging is adequate for many applications, many uncertainties remain regarding the suitability of the charge especially if the end-product is used in diverse environments. In such cases the best way of matching charger performance to the application's needs is some form of feedback control. The various types of feedback-controlled charging are discussed in this section.

The term *fast-charge* applies to charging techniques which permit the battery to be charged at the 1C rate or greater resulting in a full charge in one hour or less. Fast-charging schemes require terminating the high-rate charge before the battery receives too much overcharge. Sealed nickel-cadmium cells cannot sustain indefinite overcharge at fast-charge rates.

Most fast chargers use the split-rate constant-current approach. Split-rate charging first charges the battery at a high rate. Then, when the battery is fully charged or nearly fully charged, the charging rate switches to a slow rate. The slow rate may be 0.1C or a lower trickle rate because the battery has already reached a high state of charge.

Fast charging also has the advantage of being able to partially charge a battery in a very short time, often a matter of minutes. It is ideal for those applications requiring a number of charge/discharge cycles per day.

Obtaining fast charges through charge control requires some tradeoffs in charger design. Most charge-control schemes use a sensor which adds to circuit complexity and cost. The charger power supply must provide high charge currents which means greater cost, size, and weight relative to a standard-rate charger. There is always a possibility that a fast-charge control will fail to terminate the high-rate charge. A thermal fuse located physically in the battery and electrically in the charging path may be necessary to guard against failure of the charge-control system. The increase in battery temperature, resulting from overcharging at the fast-charge rate, opens the thermal fuse and thus terminates charging.

The following sections are devoted to the various forms of charge control employed in fast-charge systems.

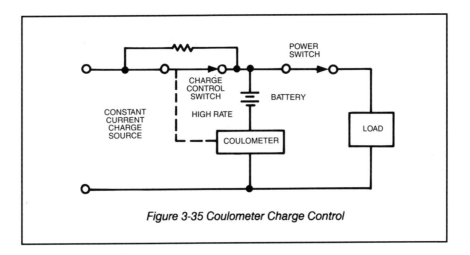

*Figure 3-35 Coulometer Charge Control*

### 3.2.6.1   Coulometric Control

The most fundamental method of fast charging is simply charging the battery until it receives just the amount of capacity needed to bring it to full charge and no more. This is done by measuring, with a coulometer, the amount of charge in ampere-minutes removed from the battery during each discharge and then measuring the amount of charge returned during each subsequent charge. The fast-charge current then ends when the number of ampere-minutes returned to the battery equals the amount needed to charge the battery.

The coulometer used for charge control is a device capable of integrating current over time and providing a signal when a predetermined current-time product has been reached. The coulometer and the battery are electrically connected in series. The coulometer can be a chemical or an electronic device. Figure 3-35 shows a block diagram of a coulometer charge control. When the battery discharges through the load, the current also passes through the coulometer. The coulometer integrates the discharge current-time product. During a subsequent charge, the coulometer integrates the charge current-time product. When the coulometer has integrated a charge equal to the previously integrated discharge it provides a signal to terminate the charge. A scaling factor is usually applied to account for the less than 100 per cent efficiency of charging.

Coulometer fast-charge control works best when the battery does not rest for long periods after charging. Since the coulometer may not possess a self-discharge characteristic similar to nickel-cadmium cells, it may not provide an accurate measure of the input required to offset self-discharge and fully charge the battery. A two-rate system compensates for this by using coulometer control of the fast-charge rate and then switching to a standard rate to ensure full charge.

### 3.2.6.2   Time Control

One of the most straightforward approaches to charge control is to use some form of timer on the high-rate charge input.

#### 3.2.6.2.1 Simple Timed Control

A simple time-limit control can be used to reduce the charge current from a high rate to a trickle rate at some predetermined period. Time-limit control can restrict the period of the initial charge rate to the amount of time required for a discharged battery to reach full charge from complete discharge, for example, 16 hours for 0.1C rate or 4 hours for 0.3C rate. This approach's liability is that the battery receives the same amount of charge without regard to the depth of the previous discharge. If the previous discharge was shallow and the battery is already near full charge, it will still receive the full length of charge. As a result the battery may be receiving high-rate current in overcharge for much of the charge time. Even so, the period of overcharge temperature stress established by the time-limit control may be small compared to the period of stress that would result from an uncontrolled charge time.

### 3.2.6.2.2 Dump-Timed Control

If the battery characteristics are known and the state of charge of the battery is known, a timed charge of extremely high rate can be safely delivered to a sealed nickel-cadmium battery that uses high-rate cells. The problem with this approach is knowing the state of charge for the battery. Otherwise the battery may spend unwanted time in overcharge receiving a very high-rate charge. This is the problem seen with simple timed control as discussed above.

One way to ensure that the cell is discharged enough to safely tolerate a timed fast charge is to first apply a discharge load that will remove the proposed charge input, even from a fully charged cell. The dump-timed charge method discharges the cell to a level that will accommodate the charge which immediately follows. The discharge given the cell prior to charge is normally at a rate much higher than the charge rate thus requiring much less time for the discharge than for the charge. This fast discharge has been termed a *dump*. When it is followed by a timed charge, the term *dump-timed charge* (DTC) is used.

The concept is best suited for single-cell battery applications. Multicell battery applications require very careful analysis of the individual cell characteristics and the interaction between cells with regard to both charge and discharge. Both multi- and single-cell applications need a cell designed for the charge rate used. The following discussion about dump-timed charge (DTC) is limited to single-cell battery applications.

Figure 3-36 shows the DTC circuit elements and the relationship described in the paragraph above. The cell is discharged at a high rate in a short time, $T_1$. The cell is then charged at a fast rate for a set period of time, $T_2$. The product of the dump time and dump rate can be expressed in ampere-hours as $Q_1$. The product of the charge time and charge rate is $Q_2$. As long as $Q_1$ is equal to or greater than $Q_2$, the times and rates can be varied to suit the application, subject to the particular cell characteristics. For example, using a cell designed for DTC, a completely discharged 1.0 ampere-hour cell in a given application may be fast charged safely at its 5C rate (5.0 amperes) for up to 10 minutes before the timer terminates the fast-charge current. With the beginning state of charge of the cell at zero, it may be charged at the fast rate for a rate-time product equivalent to almost one hundred per cent of the rated capacity of the cell. Once the fast charge is complete a trickle charge may still be necessary to top the charge and replace self-discharge losses.

### 3.2.6.3   Temperature Sensing Control

Fast-charge sealed nickel-cadmium cells and batteries may be fast charged using battery temperature as the signal to switch from the high charge current, provided that certain precautions are followed. Several methods, each with its own attributes and limitations, can be used to sense temperature and thereby control fast charging. The following subsections describe the various methods and present information to help select the best approach for the specific application being considered.

Temperature control of charging relies on the increase in cell internal pressure and subsequent increase in cell temperature as the cell approaches full charge and

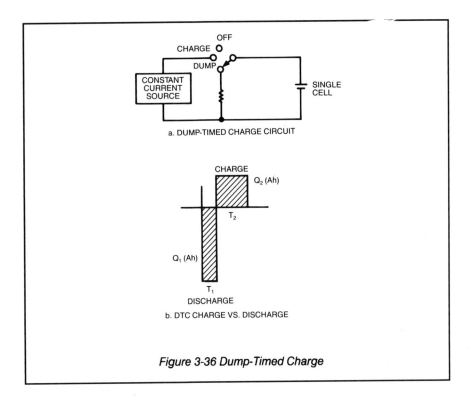

*Figure 3-36 Dump-Timed Charge*

then goes into overcharge. Both of these parameters are related to the charge rate and the temperature at which the charging is conducted.

The relationship of cell temperature and cell internal pressure to charge input is shown in Figure 3-37. The pressure begins to rise before the charge input reaches 100 per cent because some oxygen gas is generated. Consequently the charging efficiency decreases in this region (see the discussion in Section 3.2.2). The rate of

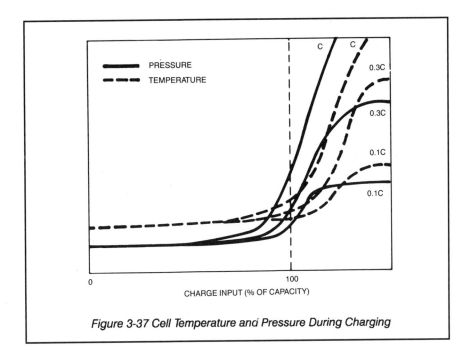

*Figure 3-37 Cell Temperature and Pressure During Charging*

pressure rise and the level that the pressure reaches are functions of the charge rate as shown by the solid line curves in Figure 3-37. The cell temperature rise also varies with the charge rate as shown by the dotted lines in Figure 3-37. The temperature rise lags the pressure rise because of the thermal inertia of the cell and because oxygen pressure must exist before oxygen recombination can cause the heat generation that increases cell temperature. The temperature rise is generally detectable when the cell approaches full charge. The amount of temperature rise for any specific cell size and cell construction has a direct relationship to charge current.

The ambient temperature around the battery during charging affects both the pressure and the temperature at which the battery will stabilize. The pressure required to achieve a given rate of oxygen reduction at the negative electrode is an inverse function of temperature. At low cell temperatures the oxygen pressure can rise rapidly and may reach the release pressure of the safety vent mechanism. At high cell temperatures the pressure rise will be substantially less than exists at normal temperatures because oxygen recombination is enhanced.

A variety of methods of using the cell's temperature rise to control the fast charge current are available. In the following sections, four control methods are discussed:

- Temperature Cutoff (TCO)
- Incremental Temperature Cutoff ( ΔTCO)
- Differential Temperature Control (ΔT Control)
- Rate of Temperature Change Control (dT/dt Control)

### 3.2.6.3.1 *Temperature Sensing Methods*

The following information on temperature sensors applies to all of the temperature cutoff concepts.

*Thermostat Temperature Sensing*    The simplest temperature sensor for charge-rate cut-off is the manual-reset thermostat. This is a switch that opens when the rising cell temperature reaches the setpoint and can be closed manually when the temperature falls below the reset temperature which is lower than the setpoint. In battery applications, the thermostat is thermally coupled to a selected cell in the battery pack and electrically connected in series with the charging circuit (See Figure 3-38). A bypass resistor around the thermostat drops the charging current to a trickle-charge rate when the thermostat is open. The combination of manual-reset thermostat and bypass resistor is a uniquely simple fast-charge system.

While the advantage of the manual-reset thermostat is its low-cost approach to sensing the cutoff point and switching the fast-charge current, it has an operational disadvantage. There is a dead band between opening and reclosing of the thermostat. Fast charge cannot be restarted if the battery temperature is above the reset temperature of the thermostat, that is, above the opening temperature or in the dead band of the thermostat. This elevated temperature might exist after the battery has been discharged at a high rate or in a high-temperature environment. The battery has to cool to the reset temperature before fast charging can be started.

If an automatic-reset thermostat is used in place of a manual-reset thermostat, an undesirable bouncing condition can occur in overcharge. The thermostat, heated to its opening (setpoint) temperature by the battery, terminates fast charge. When the battery cools, the thermostat falls below its closing (reset) temperature, and fast charge is again initiated. This repeated on-off condition, illustrated in Figure 3-39, shortens battery life because the cells are exposed to higher average temperatures and pressures.

Of course, the manual-reset thermostat avoids this on-off bouncing. It will not

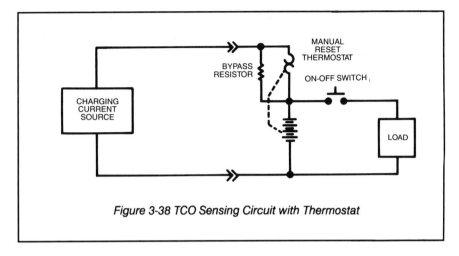

*Figure 3-38 TCO Sensing Circuit with Thermostat*

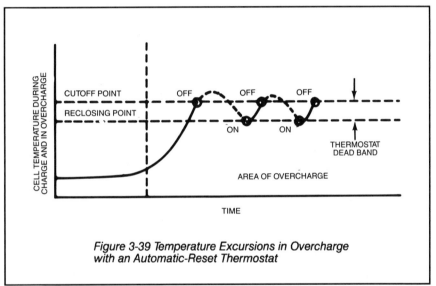

*Figure 3-39 Temperature Excursions in Overcharge with an Automatic-Reset Thermostat*

close until it is reset by pushing its reset button. A commonly used method to avoid bounce in automatic-reset thermostats is to provide an electronic function in the charger that detects the opening of the thermostat and then keeps the fast-charge current source disconnected until a positive action has taken place (such as removing the battery from the charger and then reconnecting it) which is detected by the electronic latching mechanism.

*Thermistor Temperature Sensing* Thermistors are temperature-sensitive resistors that can be used to sense the surface temperature of a cell in a battery pack. Thermistors are available with positive temperature coefficient (PTC) and negative temperature coefficients (NTC) of resistance. PTC thermistors with a large temperature coefficient in the region of interest are readily available. Because of this large temperature coefficient in a relatively narrow temperature region, this type of PTC thermistor is called a switch-type thermistor. A circuit using such a thermistor is less critical and generally need not include a costly potentiometer for establishing a setpoint. The switch-type PTC thermistor is appropriate for temperature cutoff (TCO) control of fast charge. However, the switch-type PTC does not permit matched pairs that behave well over a wide temperature range and thus NTC thermistors are used

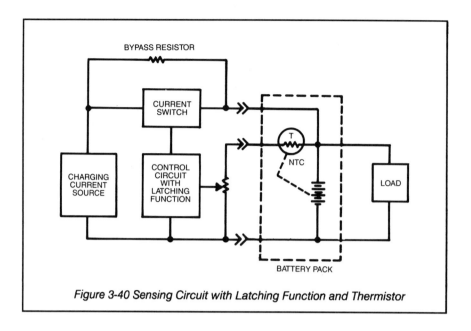

BYPASS RESISTOR

CURRENT SWITCH

CHARGING CURRENT SOURCE

CONTROL CIRCUIT WITH LATCHING FUNCTION

T
NTC

LOAD

BATTERY PACK

*Figure 3-40 Sensing Circuit with Latching Function and Thermistor*

for incremental temperature cutoff ($\Delta$TCO) fast-charge control. Switch-type PTC thermistors are typically more costly than NTC thermistors. Thermistors are signal-level devices that provide input to the charge-control electronics. Unlike thermostats, thermistors are not used as switches in the fast-charge current path.

For temperature cutoff control, the signal developed by the thermistor circuit is continuously applied to the control circuit as shown in Figure 3-40. When the temperature rises to the setpoint, the control circuit acts to open the fast-charge current switch, lowering the charge rate to the trickle rate. The latching function in the control circuit will hold the current switch open when the temperature of the cell falls thus avoiding bouncing back into high rate overcharge.

A thermistor is small enough to fit into almost any battery pack without increasing its size. The thermistor and its associated control circuit has no dead band which means that fast-charge can be initiated whenever the cell temperature is below the cutoff setting.

### 3.2.6.3.2 Temperature Cutoff

The temperature cutoff (TCO) charger using a manual-reset thermostat is the simplest of all temperature-based fast-charge control methods. In its most basic form it consists of a constant-current DC power supply and a battery containing a manual-reset thermostat that is tightly coupled thermally to the surface of one of the cells. As the fast-charge rate is applied to the battery, the battery accepts charge with only a small temperature rise until the cells approach full charge. As the cells transition into overcharge, the cell temperature rises sharply. The manual-reset thermostat opens at a temperature chosen to end the fast charge well before the cell vents. Such chargers are typically designed so that once the thermostat has opened, the charge current is not terminated but only reduced to an acceptable trickle-charge rate.

This sequence of events is represented in Figure 3-41. If the charge is started with cells at room temperature, the cell pressure and temperature show little increase until near full charge. The degree of lag between the pressure rise and the temperature rise determines the usefulness of temperature rise as a fast-charge control.

In Figure 3-41, the temperature sensor terminates the fast charge when the cell temperature reaches 45°C. At this time the gas pressure within the cell is climbing

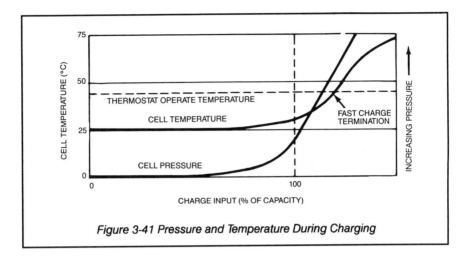

*Figure 3-41 Pressure and Temperature During Charging*

rapidly but is normally within operating limits. A fast-charge system utilizing fast-charge cells and a TCO charger will, in most circumstances, provide a perfectly viable system. However, three special situations must be considered:

- Battery Charged In A Low-Temperature Environment — Low temperatures pose a problem for the TCO concept. When a battery is charged starting from a low temperature, it may fail to reach the predetermined cutoff temperature, leading to uncontrolled fast charging and permanent cell damage. The two curves in Figure 3-42 illustrate the same battery at the same charge rate but charged in two different ambient temperatures. The battery charged at room temperature, Curve A, reaches the cutoff temperature setting. The battery in the cold environment, Curve B, fails to cutoff the fast-charge current.
- Cold Battery Placed In The Charger — This condition may also result in a serious problem for the fast-charge TCO concept. If the battery is cold at the start of charging, even if charged in a 23°C ambient temperature, the monitored cell may not reach

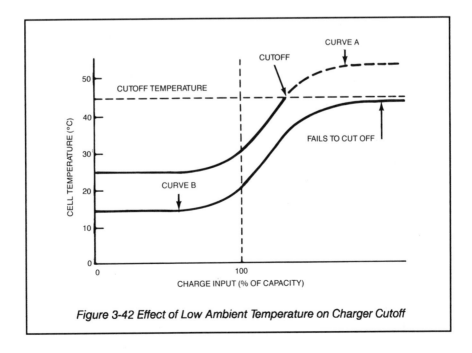

*Figure 3-42 Effect of Low Ambient Temperature on Charger Cutoff*

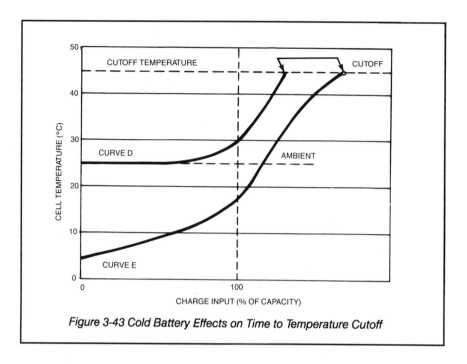

*Figure 3-43 Cold Battery Effects on Time to Temperature Cutoff*

the cutoff temperature before the pressure in the cell has risen to the point where venting is occurring. The temperature profile for a cold battery looks like Curve E in Figure 3-43. The time in overcharge to reach the temperature cutoff point for Curve E is longer than for a battery at room temperature, represented by Curve D. Therefore, the peak pressure would be significantly higher for battery E than D. This is in addition to the problems covered in Section 3.2.7.5 on low-temperature charging.

- Hot Battery Placed in the Charger — When charging is begun with the battery temperature above the setpoint of the temperature sensor, the charger will deliver only the trickle-charge rate to the battery. When a simple latch to the trickle-charge rate is used, charging must be restarted when the battery temperature has fallen below the thermostat's reset temperature. It is possible, however, to provide a circuit which will automatically begin fast charging when the battery temperature falls below the reset temperature if the battery has not previously received fast-charge current.

Except for the three situations above, the TCO concept, when used within cell specification charge rates and appropriate cutoff temperature settings, is a very reliable charging system. If either of the two cold temperature application situations is likely to occur, one of the other temperature sensing systems should be selected.

### 3.2.6.3.3 Incremental Temperature Cutoff

The Incremental Temperature Cutoff (ΔTCO or delta TCO) charger overcomes the cold environment charging problems seen with the simple TCO fast charger.The ΔTCO charger switches from fast charge when the battery temperature has risen by a predetermined increment above the ambient temperature.

Figure 3-42 shows the effect of charging a battery in a cold ambient temperature (Curve B) and at room temperature (Curve A). Using the ΔTCO concept, batteries in either environment will switch reliably after a 10°C rise. In both situations shown in Figure 3-42 the cutoff with a 10°C ΔTCO occurs comparatively early avoiding excessive fast charging thereby prolonging battery life. In the case of Curve B in Figure 3-42 the ΔTCO method cuts off while the normal TCO does not.

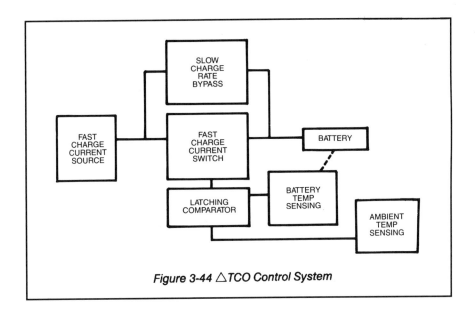

*Figure 3-44 △TCO Control System*

The ΔTCO system requires two temperature sensors. One is mounted to measure the surface temperature of the highest temperature cell in the battery (normally the centermost cell). The other sensor is thermally isolated from the heat generated by the battery and charger so it senses the temperature of the surrounding air. The charge control then switches the fast-charge current off when the battery temperature sensor reaches a preselected increment above the temperature of the ambient sensor. The choice of temperature increment (delta) varies depending upon battery type, cell size, number of cells in the battery, battery packaging, and other thermal considerations. The 10°C value is often used.

The temperature sensors are normally thermistors or thermocouples providing a signal to a comparator circuit within the charger as shown in Figure 3-44. When the temperature difference reaches the prescribed increment, the comparator circuit signals the fast-charge switch to terminate the fast-charge current. A trickle-charge current normally continues after the high-rate current is cut off.

### 3.2.6.3.4 *Differential Temperature Control*

Another method of fast-charge control using temperature rise to reduce the charge rate is the Differential Temperature (ΔT) Control Method. This charger reduces the charge current in overcharge to a rate just adequate to maintain a predetermined battery temperature rise above ambient. Figure 3-45 shows the concept. Output signals from a battery temperature sensor and an ambient temperature sensor provide inputs to a nonlatching comparator which adjusts the charge rate based on the difference between the two signals. The sensors are chosen, or their signal outputs are conditioned, so when the battery temperature has risen above the ambient temperature by the set increment, e.g. 5°C, the comparator and charge current control will produce a smooth transition of current from the fast-charge rate to an overcharge rate that just maintains the predetermined temperature rise (ΔT).

Because there is no latch, this control system automatically permits a battery, which was hot when placed in the charger, to receive the fast-charge rate when it cools to a temperature less than ΔT above the ambient temperature. This charge control method requires design precautions to prevent ambient temperature changes, caused by air conditioning and heating systems, from causing the charger to deliver pulses of fast-charge current when the battery is in overcharge. Such pulses could

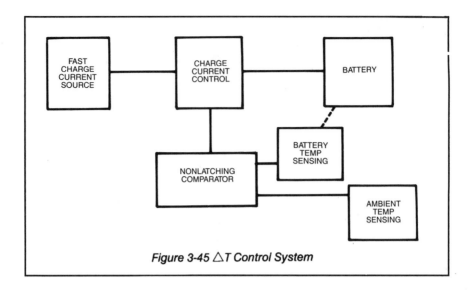

*Figure 3-45 △T Control System*

provide an overcharge rate greater than the cell specification rate, therefore proving abusive. Consequently, the method requires extensive system engineering.

### 3.2.6.3.5 *Rate of Temperature Change Control (dT/dt Control)*

The temperature versus time curve for a sealed nickel-cadmium battery receiving fast charge has the shape shown in Figure 3-46. This characteristic shape results in the rate of temperature change (dT/dt) becoming a useful value as the cell nears full charge.

The signal from a temperature sensor can be electronically differentiated with respect to time in two ways. The first is to use an operational amplifier. This method is limited by the sensitivity of the operational amplifier to electrical noise. In a noisy environment, it readily produces extraneous outputs resulting in inaccurate control.

The second and better approach is to sample the temperature on regular short intervals and then estimate dT/dt by using the temperature change $\Delta T$ based on the average of samples grouped around points in time separated by a time interval $\Delta t$. This is easily done with a microcomputer chip. Control using dT/dt becomes particu-

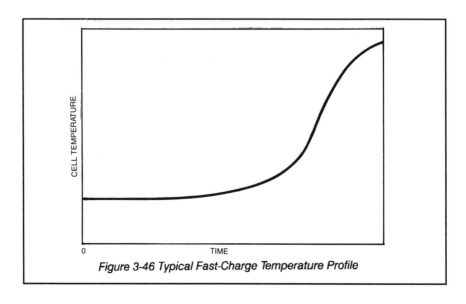

*Figure 3-46 Typical Fast-Charge Temperature Profile*

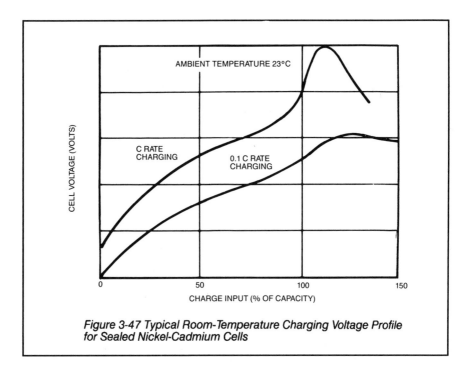

*Figure 3-47 Typical Room-Temperature Charging Voltage Profile for Sealed Nickel-Cadmium Cells*

larly attractive if the microcomputer chip is already present for some other purpose. The microcomputer approach requires analog-to-digital (A/D) conversion of temperature sensor signals to provide usable inputs to the microcomputer. The change in dT/dt that the microcomputer can detect (and thus how early in the temperature rise fast charge can be terminated) is dependent on the resolution and accuracy of the A/D converter. Here, of course, noise is again a factor.

### 3.2.6.4   Voltage Sensing Control

Battery voltage can be used as a signal for controlling charge current, but the many factors affecting nickel-cadmium cell charging voltage such as temperature, age of cell, previous history, and type of construction must be taken into account. These variables can, in many instances, have a greater effect on battery voltage than the state of charge. Therefore, voltage sensing for charge control must consider the points described below.

A primary failure mode for nickel-cadmium cells at the end of their life is internal shorting as described in Section 3.4. **Any charging method that depends solely on voltage magnitude to terminate fast charge may lose control when a cell shorts.** Voltage cutoff (VCO) and constant potential (CP) charging are examples of charge control methods that are subject to this problem. Loss of control means the battery may be fast charged while it is fully charged with resulting venting and impact on battery life.

The voltage of a fully discharged sealed nickel-cadmium cell follows a profile similar to that shown in Figure 3-47 during charging. At room temperature a fully discharged fast-charge cell, when placed on charge at the 1C rate, rapidly rises to about 1.40 volts and then gradually climbs to about 1.45 to 1.50 volts as the cell approaches full charge. At this time, there is little uncharged positive plate active material remaining and oxygen is being generated at the positive plate, the cell voltage again rises somewhat more rapidly. Once the cell temperature is increasing rapidly, the voltage will begin to decrease.

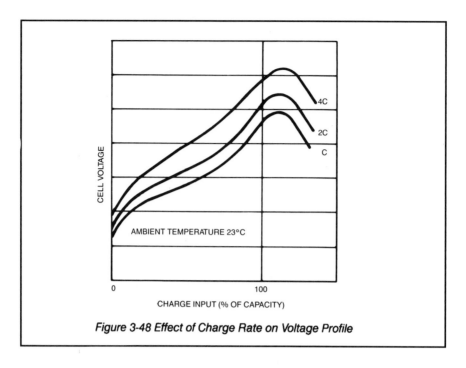

*Figure 3-48 Effect of Charge Rate on Voltage Profile*

The peak cell voltage and the rate of rise depend on the charge rate, as shown in Figure 3-48. The degree of the cell voltage rise upon approaching full charge is also dependent on other factors such as the cell design, age, and previous history. Some cells produce a pronounced voltage rise and other cells produce a minimal voltage rise. The curves shown in the figures are general representations of typical fast-charge cells.

The charge voltage of all sealed nickel-cadmium cells varies with the temperature of the cell. If the cell is very cold the charge voltage can rise to a high value. Conversely, the voltage of a warm cell during charging will be lower, as shown in Figure 3-49.

Sealed nickel-cadmium cells exhibit a charging voltage which varies inversely with temperature by about 3 millivolts per degree Celsius. The charge voltage of sealed nickel-cadmium cells also varies from cell type to cell type depending upon the design of the cell. These differences in design, construction, and manufacturing processing affect cell charging voltage. An indication of the range of voltages is shown in Figure 3-50. The three curves shown are representative of charging voltage differences which may be expected from cells designed or manufactured differently.

It is important to appreciate that different charging voltage characteristics will be exhibited by a battery depending upon:

- Charge rate as the battery approaches full charge
- Cell temperature during charging
- Cell design and construction

For this reason, the product designer should consult with the cell manufacturer when designing a fast-charge control based on voltage. The charge rate, setpoint, charger design, and degree of temperature compensation for the signal must be selected in light of the cell type. A voltage-sensing control system may lead to product field problems if battery characteristics and variations are not understood and taken into consideration. **Given the possible difficulties with charge control based on the magnitude of the voltage, it is generally not recommended for routine application.**

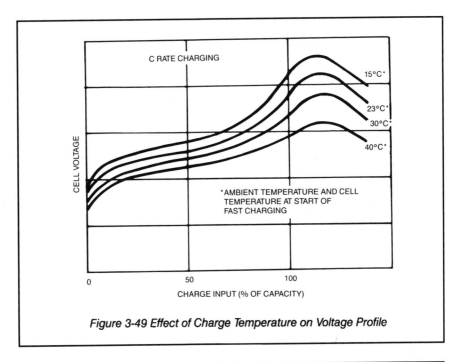

*Figure 3-49 Effect of Charge Temperature on Voltage Profile*

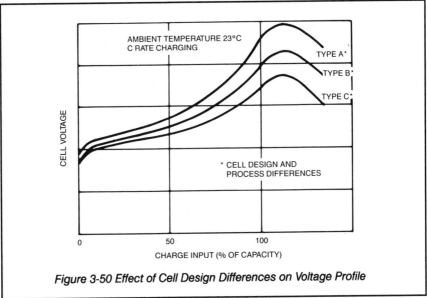

*Figure 3-50 Effect of Cell Design Differences on Voltage Profile*

### 3.2.6.4.1 Constant-Potential Charging

A constant-potential (CP) charging source implies that the charger maintains a constant voltage independent of the charge current load. Realistically most CP-type battery chargers have limitations on the current they can supply.

Charging from a CP source is quite common for vented nickel-cadmium and lead-acid batteries. It is rarely used for sealed nickel-cadmium batteries because their charging characteristics are different. Figure 3-51 illustrates this difference in terms of voltage response to a constant charge current. Vented nickel-cadmium and lead-acid batteries exhibit a pronounced voltage increase in response to an applied constant current as the battery approaches full charge. This is caused by the onset of

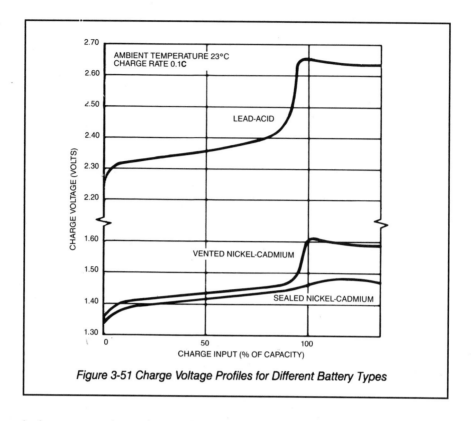

*Figure 3-51 Charge Voltage Profiles for Different Battery Types*

hydrogen generation at the negative plate. In sealed nickel-cadmium cells, the voltage rise is relatively small. The sealed cell's negative plate is depolarized by the recombination of oxygen so that little hydrogen overvoltage or hydrogen gas generation occurs.

**Constant-potential charging of sealed nickel-cadmium batteries is strongly discouraged by Gates Energy Products.**

Among the reasons for discouraging use of constant-potential charging is the problem of *thermal runaway*. When charged at 23°C, the voltage of a sealed nickel-cadmium battery must reach about 1.45 volts per cell in order to achieve full charge. If charging is done from a CP source of 1.45 volts, the overcharge current would be about 0.1C if the cell temperature remained constant.

The heat generated from an initial 0.1C overcharge current can cause a significant temperature rise for sealed nickel-cadmium batteries. The related increase in electrochemical activity reduces the electrochemical polarization. Maintaining a fixed charging voltage as the temperature rises results in an overcharge current greater than would have occurred had the nickel-cadmium battery remained at the ambient temperature.

If heat transfer from the sealed nickel-cadmium battery is less than excellent, or if the ambient temperature increases, overcharge current may continue to escalate with an attendant increase in battery temperature. This system instability associated with a fixed-voltage charging source is *thermal runaway*. The severity of the runaway resulting from this loss of charge control is determined by the current limit of the charger.

### 3.2.6.4.2 *Voltage Cutoff*

The voltage-cutoff (VCO) control method uses an increase in voltage magnitude as the signal to terminate the fast charge. As shown in Figure 3-52, the voltage cutoff

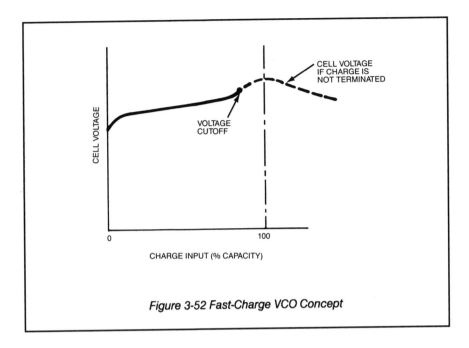

*Figure 3-52 Fast-Charge VCO Concept*

method senses the increase in battery voltage magnitude as the battery approaches full charge. When the voltage rises to the predetermined voltage point, the charger circuit switches off the fast-charge current.

Normally, a trickle-charge current continues to the battery once the fast-charge current is terminated. The charger circuit to accomplish this switching is illustrated by the block diagram of Figure 3-53. The VCO level must be automatically adjusted

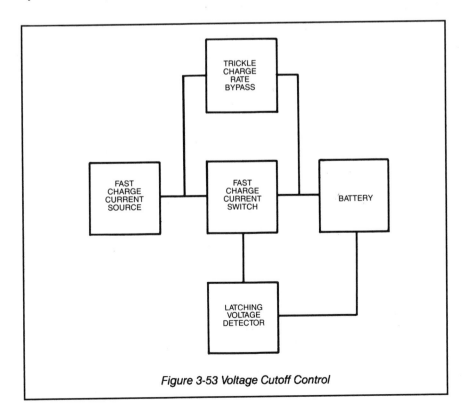

*Figure 3-53 Voltage Cutoff Control*

for battery temperature because of the temperature dependence of the battery charge voltage.

If the cutoff point is not automatically adjusted for cell temperature, one of the following could occur:

1. Cold Cells in a Cold Ambient — The battery voltage will reach the normal cutoff voltage point quite early. If the cutoff value is not automatically adjusted for cell temperature, the battery will receive only a partial charge before cutoff.
2. Warm Cells in a Warm Ambient—The battery voltage depressed by high temperature may never reach the cutoff voltage point. If the cutoff value is not adjusted for cell temperature, the battery will continue to charge at the fast-charge rate, leading to permanent cell damage and a reduction in life.

**Charge control by sensing the absolute voltage rise is strongly discouraged by Gates Energy Products.**

### 3.2.6.4.3 Rate-of-Voltage Change

Rate-of-voltage-change (dV/dt) charge control limits overcharge by terminating fast charging at some point where the voltage begins to rise prior to Point 2 in Zone C of Figure 3-54. The relatively large magnitude of dV/dt in this part of the voltage profile presents an opportunity for terminating fast charge prior to the voltage maximum. The dV/dt charge control system must be designed to ignore the relatively large dV/dt in Zone A of the charge voltage characteristic. The system may be designed not to respond to a dV/dt for some initial period or not to respond until the battery voltage has exceeded some voltage plateau such as at Point 1 in Zone B.

The ability to terminate fast-charge current on the basis of dV/dt depends on a reliable and repeatable voltage profile. In certain circumstances the battery may not always provide the expected voltage profile and this could result in the fast-charge current not being cutoff. This means a backup termination method like a timer is required.

A microcomputer chip is technologically the best approach to process the dV/dt signal. An analog-to-digital (A/D) converter is required to provide a usable voltage signal to the microcomputer.

### 3.2.6.4.4 Inflection Point Cutoff

There is usually an inflection point ($d^2V/dt^2 = 0$) near the midpoint of Zone C of Figure 3-54 if the battery has been fully discharged or nearly fully discharged prior to charging. That is, the slope of the voltage profile at this point transitions from an increasing positive value to a decreasing positive value. Using this feature to accomplish an inflection point cutoff (IPCO) of fast charge has appeal because it precedes the voltage peak, limits the amount of fast charging as full charge is approached, and requires only two terminal connections to the battery.

At the inflection point, the second derivative of voltage is zero. Because a zero value of second derivative is rather difficult to detect as such, another approach is to sense the decreasing value of positive dV/dt ($d^2V/dt^2 < 0$) which occurs just beyond the inflection point. The profile in Zone A of Figure 3-54 also has a decreasing slope. This complication requires a charger design that will not respond to a decreasing slope until: (1) after a prescribed time delay, (2) after a predetermined voltage has been reached, or (3) after some other enabling procedure has been completed.

There are conditions where a battery may not display a characteristic that will meet the criteria established for fast-charge cutoff based upon the inflection point. Applying a fast-charge to a fully charged battery is an example. Therefore, a backup

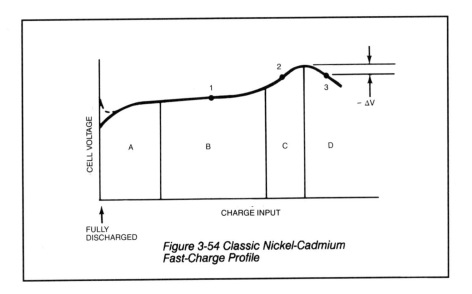

*Figure 3-54 Classic Nickel-Cadmium Fast-Charge Profile*

cutoff method (such as –ΔV discussed below) is required for this type of fast-charge control.

### 3.2.6.4.5 *Voltage-Decrement Cutoff*

As shown in Figure 3-54, a typical battery during fast charging will exhibit a voltage maximum at the transition from Zone C to D, followed by a voltage decrease resulting from battery temperature rise. At moderate charging temperatures the decrease in voltage beyond the peak is well defined. It can be detected by an analog circuit or a microcomputer chip which remembers the highest voltage it has seen and responds to a voltage decrement (–ΔV) relative to the maximum. Voltage-decrement cutoff (VDCO) charge control requires only two terminal connections. A usable voltage-decrement will typically produce a battery stress equivalent to a ΔTCO charge control of about 5 to 10°C.

At elevated battery temperatures the negative voltage slope beyond the charge voltage peak is reduced, causing the VDCO to be delayed relative to normal operating temperatures for a given –ΔV detection sensitivity.

This delay is generally not a problem unless the sensitivity of the control circuit is such as to require a –ΔV greater than about -30mV/cell and the battery temperature exceeds 40°C.

### 3.2.6.5   **Voltage and Temperature Sensing Control**

One of the most reliable fast-charge concepts is the use of both voltage and temperature signals to sense when the fast-charge current should be terminated. Two concepts using voltage and temperature of the battery are described below.

### 3.2.6.5.1 *Voltage-Temperature Cutoff*

The earlier discussions about temperature-sensing charge controls and voltage-sensing controls for fast charging sealed nickel-cadmium batteries identified some limitations of each control system. A very reliable fast-charge system is obtained by using both the voltage and temperature of the battery in an OR logic form to act as indicators of approaching full charge.

The voltage-temperature cutoff (VTCO) charger senses both voltage and temperature. The charger is designed to monitor battery voltage and the temperature of one or more cells in the battery. Fast-charge current is terminated when the battery voltage or temperature exceeds a predetermined level.

The voltage level used is typically the voltage of the entire battery. The temperature of a cell (or of a number of cells) is monitored by a temperature-sensing device. The precise voltages and temperatures used for cutoff depend upon the type of cells in the battery, the application temperature range, and the fast-charge current. With this control method, both temperature and voltage values can be set to achieve a high state of charge before the fast-charge current is terminated, without compromising battery life or reliability.

### 3.2.6.5.2 *Voltage Limit Temperature Cutoff*

Voltage Limit Temperature Cutoff (VLTCO) is simply a variation of VTCO. The VTCO avoids hydrogen generation by preventing the battery voltage from exceeding a given level, but, at low temperatures, VTCO can result in premature fast-charge cutoff with a resulting decrease in delivered capacity. A voltage limit, instead of a cutoff based on voltage, can provide the same protection against hydrogen generation in low temperature applications without resulting in premature cutoff. The voltage limit, when charging a cold battery, causes a decrease in charge current until battery temperature increases as full charge is approached. At that time increasing cell temperature will cause an increase in charge current, driving the battery to TCO without having exceeded a critical voltage along the way. It is imperative, however, that the fast-charge current be sufficient to drive the battery temperature to the TCO level at the lowest required ambient charging temperature or the fast charge will not be terminated.

With VLTCO control and a given combination of battery heat transfer characteristic and ambient temperature, it is possible for battery temperature to stabilize in the fast-charge mode at a temperature below TCO, even if the fast-charge current would normally drive the battery temperature to TCO. This is not a likely event, but as a precaution the design should be checked for this possibility before the VLTCO charge control method is implemented.

## 3.2.7 CHARGING POWER SOURCES

Sealed nickel-cadmium cells and batteries may be charged from either a direct current (DC) power source (such as another battery, a DC generator, or a photovoltaic cell array), or they can be charged from a rectified alternating current (AC) power source. Both will be discussed below. Whichever power source is chosen, sealed nickel-cadmium cells are normally charged with a relatively constant current.

The ideal constant-current source has an effective source impedance which is infinite. Approaching this ideal requires electronic regulation of charge current which is often not economically justifiable. The constant-current charger, in practice, has a source impedance just large enough to acceptably restrict overcharge current variations resulting from source and battery voltage variations.

### 3.2.7.1 DC Power Source

A very simple method of charging utilizes a DC power source as shown in Figure 3-55. The charge current is calculated by the equation below:

$$I_{ch} = (E_s - E_b)/R$$

Where:  $I_{ch}$ = charge current
$E_s$ = source voltage

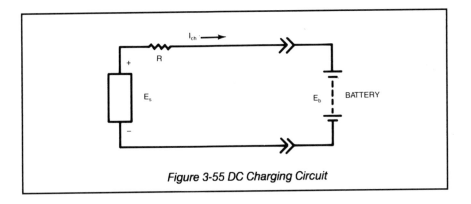

*Figure 3-55 DC Charging Circuit*

$E_b$ = battery charge voltage (1.45V/cell typical)

R = current-limiting resistance

If the source voltage is only slightly greater than the battery charge voltage, small source voltage or battery voltage variations result in large variations in charge current. If a regulated source voltage is used, its voltage may need to be only slightly greater than the battery charge voltage. However, under this condition, temperature changes in the battery may cause enough variation of the battery electrochemical polarization to result in a significant change in the charge current. (Voltage versus temperature is shown in the Tafel curves in Figure 3-30 or in the Appendix.) In batteries as opposed to cells, there is another cause for concern. If a cell within the battery shorts, the battery voltage may be substantially reduced. If the source voltage is selected to be too close to the nominal battery voltage, the result can be large charge currents through the battery. Thus it is better, whenever possible, to select a higher source voltage and a large impedance. This provides a more uniform charge current flow despite major changes in the battery voltage.

## 3.2.7.2   AC Power Source

Most constant-current chargers for nickel-cadmium batteries use the alternating current (AC) line as a power source. The basic requirements of common AC source chargers are like those of the DC power source, namely, the source voltage must be significantly greater than the battery voltage and the impedance must limit the current. The impedance of an AC power source can be reactive, resistive, or a combination of these. Limiting the current by reactance has the advantage that the reactive impedance component does not contribute to heat generation. Rectification is used to convert the alternating current into direct current that is only in the charging direction through the battery. Most chargers utilize a transformer which permits optimized matching of the source voltage to the charge voltage of the battery. Another very important feature of the transformer is that it isolates the charging output from the AC line, thus avoiding the hazard of electrical shock.

Several transformer circuits are conventionally used for charging from AC power. Those considered here are shown in Figures 3-56 through 3-58. In these circuits, current limiting is assumed to be achieved primarily by resistance. This resistance is usually accomplished by proper sizing of the wire in the transformer windings, with contribution from both the primary and secondary windings.

The nickel-cadmium battery can be charged equally well with either half-wave or full-wave current. Half-wave rectification (Figure 3-56) is the most economical for low power levels and hence is the most popular method. For higher power chargers, it is usually more economical to use full-wave rectification (Figures 3-57 and 3-58).

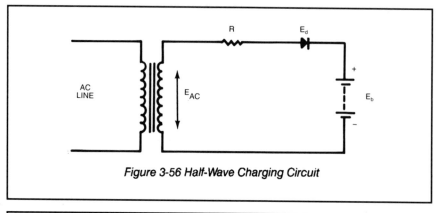

*Figure 3-56 Half-Wave Charging Circuit*

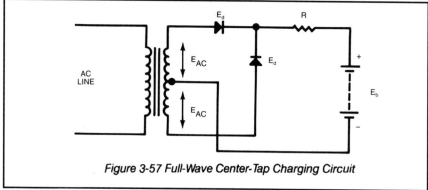

*Figure 3-57 Full-Wave Center-Tap Charging Circuit*

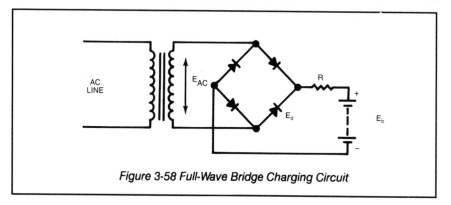

*Figure 3-58 Full-Wave Bridge Charging Circuit*

An alternative charging circuit is shown in Figure 3-59. It uses a capacitor to limit current in conjunction with a full-wave bridge rectifier. The primary advantage of this circuit is that it is essentially loss-free. With a 120V, 60 Hz source, the charge current is approximately 40 milliamperes (mA) per microfarad of capacitance. For a 220V, 50 Hz source, the charge current is approximately 60 mA per microfarad of capacitance. The capacitor must be able to withstand peak line voltage.

The peak inverse-voltage rating of the bridge rectifier in Figure 3-59 needs to equal only the battery charge voltage if the battery is never disconnected from the charge circuit, but must withstand line voltage if the battery is disconnected. The circuit is typically used for relatively low charge rates. Its primary disadvantage is that it provides no electrical isolation from the AC line. Therefore, safety considerations limit the application of capacitor-type chargers to devices where the charger and battery are completely enclosed. The enclosure must prevent user contact with

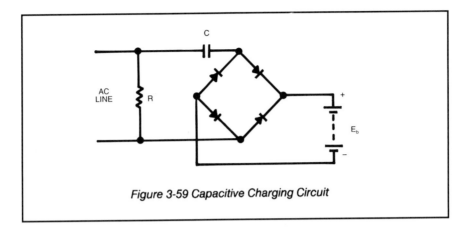

*Figure 3-59 Capacitive Charging Circuit*

any part of the circuit including the battery contacts. Resistor R in Figure 3-59 serves only to discharge the capacitor to prevent possible electrical shock at the line cord plug when the circuit is disconnected from the AC receptacle.

### 3.2.7.3   Current Regulation

The charger must provide sufficient current regulation so that the battery is not damaged when it is left in overcharge. In Section 3.2.4.1 the overcharge characteristic (Tafel curves, Figure 3-30 and in Appendix A) of sealed cells shows that cell charge voltage is a function of cell temperature as well as overcharge current. The nickel-cadmium charger current must be sufficiently well regulated to prevent the overcharge current from reaching unacceptable levels as a consequence of high battery temperature and high AC-line voltage. The overcharge rate should never exceed the cell specification charge rate by more than 20 per cent.

   To achieve an acceptably well-regulated constant-current charger design, the transformer secondary voltage of an AC power source is selected so that its open circuit voltage is about 2.2 root mean square (rms) volts per cell. The voltage drop through each series rectifying diode must also be considered. The following information may be a useful guideline for the design of simple chargers.

   The DC overcharge current with rectified AC source can be calculated using two equations:

Half-Wave:
$$I_{oc} = (\sqrt{2E_{ac}} \times K_1)/\pi R$$
Full-Wave:
$$I_{oc} = (2\sqrt{2E_{ac}} \times K_1)/\pi R$$

Where:   $I_{oc}$ = overcharge current
   $E_{ac}$ = AC rms open-circuit source (secondary) voltage (outside to center tap when center tap is used)
   R = sum of total effective transformer resistance and external resistance
   $K_1$ = coefficient from Figure 3-60
   $E_b$ = battery voltage in overcharge (used in Figure 3-60)
   $E_d$ = diode voltage drop (used in Figure 3-60)

   Overcharge current $I_{oc}$ is not a linear function of the difference between $E_{ac}$ and ($E_b + E_d$). This difference is accounted for in coefficient $K_1$ (determined from Figure 3-60). The rectifier voltage drop $E_d$ is a nonlinear function of current, but 0.8 volt is a typical value for the silicon diode; a value of 1.6 volts must be used for a silicon bridge rectifier.

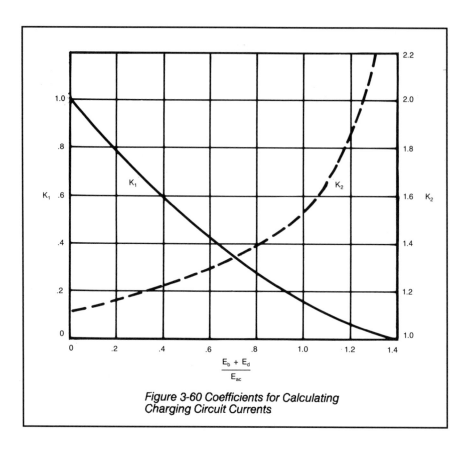

*Figure 3-60 Coefficients for Calculating Charging Circuit Currents*

The value of voltage to be entered as $E_b$ is typically 1.45 volts per cell at 23°C (see the Tafel curves in Appendix A) for 0.1C to 0.3C charge rates. The ratio of rms current to DC charging current is always greater than 1.0 in a rectified AC charging circuit. This ratio increases as the ratio of $E_{ac}$ to $(E_b + E_d)$ decreases. The rms current must be calculated to determine the ratings for the transformer, diode(s), and any external resistor. This rms current can be calculated using the appropriate equation below. Coefficient $K_2$ is also determined from Figure 3-60.

$$\text{Half-Wave: } I_{rms} = \sqrt{2}K_2I_{oc}$$
$$\text{Full-Wave: } I_{rms} = K_2I_{oc}$$

The greatest rms current will occur with high line voltage at the start of charge where battery voltage may be about 1.30 volts per cell. Charging-circuit components must be rated to tolerate these conditions for a significant portion of the charge time. These components may see even more severe conditions for a brief period if a battery has been deeply discharged to zero volts.

The stress on charging circuit components may become even more severe should a cell within a battery become shorted. It is not uncommon for the relationship between charger AC source voltage and battery voltage to be such that a ± 10 per cent AC line voltage variation will result in a ± 25 per cent variation in overcharge current. With a line voltage which is 110 per cent of nominal and with the battery at the beginning of charge, the rms current will be 135 per cent of the nominal rms current experienced in overcharge. If, in addition, one cell in a four-cell battery is shorted, the rms current will increase to 160 per cent of the normal overcharge value. For this case, the heat generated in a resistive current-limiting element

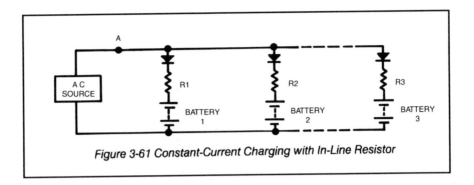

*Figure 3-61 Constant-Current Charging with In-Line Resistor*

will be $(1.60)^2$ or 2.56 times greater than that generated under normal overcharge conditions. If the current-limiting element is a simple resistor, it must be capable of dissipating this significantly greater heat load if the charging circuit is to survive a battery with a shorted cell.

### 3.2.7.4    Series/Parallel Charging

Charging of sealed nickel-cadmium cells and batteries is quite simple when the batteries or cells are connected in series. **Sealed-cell batteries should not be charged in parallel.** Because of the slight differences in the Tafel curve (Figure 3-30) of each cell, one battery in a parallel-connected group of batteries may accept more current in overcharge than the other battery. This battery heats up due to its acceptance of more current, its electrochemical polarization decreases, and it accepts even more charge current. A slight imbalance in charge acceptance can result in one battery receiving overcharge current at a level significantly greater than its cell specification rating.

It is possible to constant-current charge multiple sealed-cell nickel-cadmium batteries from a common power source, but a current-limiting resistor must be associated with each battery as indicated in Figure 3-61. If the batteries do not all have the same number of cells the source voltage must be adequate for the battery having the greatest number of cells and each resistor must be of the proper value for the number of cells and capacity in that charging branch. When an AC source is used, and all batteries have the same number of cells, a single rectifier of adequate rating can be used at point A of Figure 3-61 in place of individual rectifiers in each charging branch.

Standby applications sometimes need to have a large bank of parallel-connected batteries under continuous charge. Figure 3-62 gives a typical schematic. The batteries are independently charged through their respective resistors from the rectified source and are isolated from each other by rectifiers CR1 through CRN. When switch $S_1$ is closed, all the batteries will contribute load current equally.

### 3.2.7.5    Low-Temperature Charging

Charging at low temperatures introduces a phenomenon not observed at room temperature. At low temperatures the efficiency of the oxygen recombination reaction at the negative electrode is reduced. Depending upon the temperature and the charge rate, hydrogen gas may also be generated at the negative electrode. The result can be an increase in pressure when charging at very low temperatures. This pressure increase could be enough to cause the cell to vent.

In practice, most charging occurs at room temperature well above the temperature at which hydrogen gas generation would become a factor. In quick-charge applications when charging below 10°C or standard-charge rate applications below 0°C

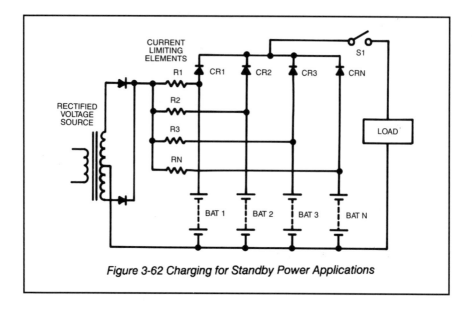

*Figure 3-62 Charging for Standby Power Applications*

the charger should be designed so that the charging current is reduced as the temperature drops, as in Figure 3-63.

The charging voltage of a cold cell will rise, not only in overcharge as shown in Figure 3-30 (the Tafel Curves), but also during the initial stages of charging. This is caused by the reduced mobility of the electrolyte ions at lower temperatures and increased electrochemical polarization at the electrodes. The charging voltage of a very cold cell will rise above 1.50 volts at the 0.1C charge rate. A charge current which brings the charge voltage of a cold cell to no greater than 1.50 volts is low enough to limit excessive gas generation. Accordingly, a voltage clamp on the charger circuit which limits charger output voltage (battery voltage) to 1.50 volts DC per cell in addition to the normal constant-current regulation is a very effective method of achieving system reliability at low temperatures.

### 3.2.7.6   Elevated Temperature Charging

As discussed earlier in Section 3.2.2.1, the ability of a sealed nickel-cadmium cell to charge effectively is reduced at elevated temperatures. The amount of charge utilized by a hot cell depends on the charge rate and its temperature. At very low charge rates (less than .05C) and cell temperature exceeding 40°C, the loss of dischargeable capacity becomes apparent. To achieve the highest level of charging effectiveness when charging above 40°C, the charge rate should be at the maximum overcharge rate allowed by the cell specifications.

But charging cells in an elevated-temperature environment at a higher rate introduces another problem. The cell temperature in overcharge rises above its ambient temperature in proportion to the charge rate. A high overcharge rate will therefore bring about a corresponding increase in cell temperature. If the ambient temperature is already high, the cell temperature in overcharge can rise to such a level that continuous overcharge will significantly reduce cell life. One way to overcome the high temperature in overcharge is to use the charge control methods discussed in Section 3.2.6.3.

### 3.2.7.7   Photovoltaic Sources

Photovoltaic cells, sometimes referred to as solar cells, are commonly used for charging nickel-cadmium batteries in space vehicles and in remote locations.

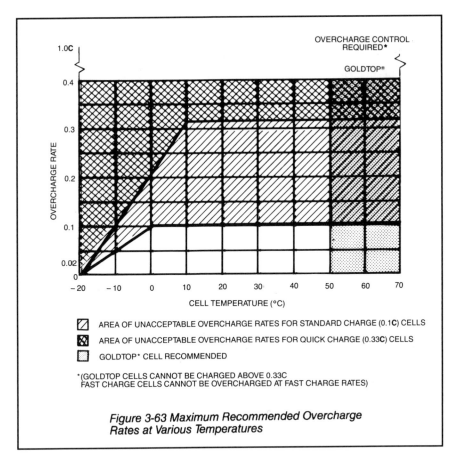

*Figure 3-63 Maximum Recommended Overcharge Rates at Various Temperatures*

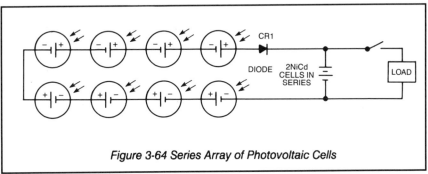

*Figure 3-64 Series Array of Photovoltaic Cells*

Figure 3-64 shows a photovoltaic array used as a nickel-cadmium battery charger. The reverse blocking diode CR1 assures that the battery will not discharge through the solar cells when light intensity is low.

Silicon photovoltaic cells are capable of delivering about 450mV when irradiated by the sun at its zenith (peak sun). There must be a sufficient number of solar cells to provide a minimum array voltage of 1.45V per nickel-cadmium cell plus the forward voltage drop of the blocking diode. Within the limits of their voltage output capability, photovoltaic cells deliver an essentially constant current which is proportional to the incident sunlight. Therefore, no current limiting resistor is required. The photovoltaic cell size (area) is typically selected to deliver rated nickel-cadmium charge current under peak sun conditions.

## 3.2.8   SUMMARY

There are a variety of charge methods which are capable of effectively charging nickel-cadmium sealed cells. These methods all accomplish the same end: putting enough energy into the battery to provide the required amount of discharge energy. The methods differ in the time it takes to accomplish the task, the complexity of circuitry used, and the techniques utilized to provide the charging function in a safe and reliable manner.

# 3.3   Storage

Nickel-cadmium cells can be successfully stored under a wide range of temperature and humidity conditions. Since some level of storage is part of virtually every battery application it is important to understand the effects of storage on both battery performance and battery life.

Storage effects can be divided into two classes: short-term effects relating to the loss of charge from a charged battery and long-term effects that may decrease battery life if the battery is improperly stored.

## 3.3.1   SHORT-TERM STORAGE EFFECTS (SELF-DISCHARGE)

All charged batteries lose their charge over time whether they are used or not. This loss of charge through internal effects is called *self-discharge*. The self-discharge rate is a function of battery chemistry and the temperature under which the battery is stored. For certain battery chemistries, such as those used in primary cells, the self-discharge rate is low: the batteries retain their charge over long periods of disuse. Most secondary batteries, including nickel-cadmium cells, have a more rapid rate of self-discharge that is compensated by the ability to recharge. In most applications, loss of charge on stand is not a major problem. However, for certain uses, prediction of capacity remaining after storage under specified conditions may be important. The following sections explain the capacity loss mechanisms and offer methods to calculate the degree of capacity lost during storage.

Note that there is a difference in storage behavior between sealed nickel-cadmium and sealed-lead cells. The sealed nickel-cadmium cell has a faster self-discharge rate than the sealed-lead battery, but it is more tolerant of storage in the discharged state.

### 3.3.1.1   Self-Discharge Mechanisms

The normal process of self discharge is the result of two electrochemical mechanisms:

1) The charged nickel hydroxide oxidizes water at electrode sites having low oxygen overvoltages. The oxygen so generated diffuses to the cadmium electrode, where it is reduced by cadmium. The net result is a loss of charge at both electrodes.

2) The second electrochemical mechanism causing self discharge is parasitic side reactions. The best examples in sealed nickel-cadmium cells are the reactions due to the naturally occurring process nitrates. The nitrate ions are reduced to ammonia at the cadmium electrode. They subsequently diffuse to the positive nickel-hydroxide electrode where they are oxidized back to nitrate ions. This nitrate shuttle cycle continues until one or both electrodes are fully discharged. With the passage of time and the use of the cell, some of the ammonia is oxidized to nitrogen and the self discharge rate is reduced, thereby improving the capacity retention characteristics of the cell.

It is for this latter reason (the oxidation of the ammonia to nitrogen) that measurement of the retained capacity performance of nickel-cadmium cells should be undertaken after the evaluation of other characteristics of importance. The resulting measurement will be more representative of the real use of the cell as well as more repeatable and accurate.

### 3.3.1.2    Retained Capacity Calculation

The rate of self discharge due to both electrochemical discharge mechanisms listed above is a function of the amount of capacity (charge) remaining in the cell and therefore tends to be exponential in form. Thus, when both the charge in the cell and the self discharge rate approach zero, the rate of decay in cell voltage also approaches zero and the voltage tends to remain in the 1.0 to 1.1 volt region for an extended period of time.

The equation for the normal capacity retention function is:

$$C_R = C_A \times e^{-t/\tau}$$

Where:    $C_R$ = retained capacity
$C_A$ = actual or initial capacity
$\tau$ = time constant in days (to 36.8% of $C_A$)
$t$ = open circuit rest time in days

The result of this equation is the typical exponential decay curve shown in Figure 3-65. The time constant for the loss of capacity is a function of both storage temperature (discussed in the following section) and cell construction. Typical time constants for sealed nickel-cadmium cells range between 100 and 200 days at room temperature.

In order to encompass the completely general case of capacity retention, a second term may be added to the exponential equation to make it:

$$C_R = C_A \times ([e^{-t/\tau}] - [t/t_r])$$

Where:    $t_r$ = number of days to 100 per cent deplete the charge through an existing shunt resistance

The shunt resistance may be either internal or external, such as an actual applied load. This equation is useful for calculating the capacity remaining in a very low-rate, long-term discharge.

### 3.3.1.3    Temperature Effects on Self-Discharge

Since the self-discharge mechanisms are the result of electrochemical reactions, they are also accelerated by increasing the temperature of the cell. The relationship between self-discharge rate and temperature is an Arrhenius function in which the logarithm of the self-discharge time constant decreases linearly with the reciprocal of the increasing cell absolute temperature as measured in degrees Kelvin.

The effect of increased temperature on retained capacity is to significantly shorten the time a cell may be stored before major loss of capacity occurs. This effect is illustrated in Figure 3-65 where both the curves for room temperature and for storage at 50°C are plotted. A very crude rule of thumb is that every decade increase in storage temperature halves the time that a battery may be stored before it loses the same amount of capacity. Better definition of temperature and design effects on retained capacity is provided on a retained capacity plot such as that included in Appendix A.

### 3.3.2    LONG-TERM STORAGE EFFECTS

Long-term storage is typically defined as storage where battery capacity has effectively dissipated. With sealed nickel-cadmium cells, this means open-circuit storage over about six months at room temperature. Although nickel-cadmium cells

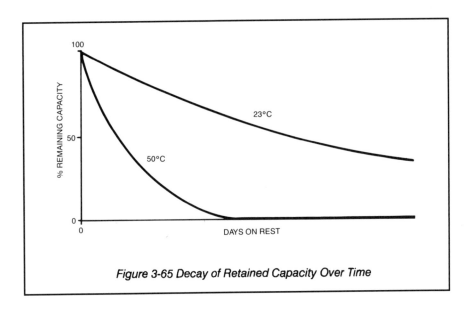

*Figure 3-65 Decay of Retained Capacity Over Time*

store very well in the open-circuit mode with an indefinite shelf life, some additional considerations in long-term storage are described below.

### 3.3.2.1 Temperature Effects During Battery Storage

Sealed nickel-cadmium batteries may be stored under a wide range of temperatures; however, storage for long periods at the upper temperature extremes may reduce battery life. To ensure the maximum life capability, long exposure to extremes of the storage temperature range should be avoided. The recommended storage is 0°C to 30°C, but cells can tolerate excursions of –40 to +100°C.

Figure 3-66 depicts the number of cycles required for full capacity recovery for cells stored for one year at various temperatures. Capacity recovery cycles increase with higher storage temperatures since the active material becomes passivated over time.

Fast-charge cells, charged at 1C or higher, will exhibit high charge voltage on the first cycle after prolonged storage at room temperature or if stored in warm temperatures. This high charge voltage may prematurely terminate fast charge in a voltage-controlled charging system.

### 3.3.2.2 Loaded-Storage Effects

One application for nickel-cadmium batteries is to provide back-up energy for volatile semiconductor memories in a wide variety of electronic devices. Although the drain rate of a typical semiconductor memory and its associated circuitry is low, it is nevertheless sufficient to discharge the battery if there are long periods of storage with the load connected to the battery. The elapsed time from battery installation in the device to powering up the device in the field may combine with self-discharge to exceed the discharge capacity of the battery. The result in this case is that an electrical load is connected to the battery while it is in the discharged state. This is referred to as *loaded storage*.

When a nickel-cadmium battery has been completely discharged but a load continues to be applied, the battery voltage will approach zero. When its voltage is held at zero or close to zero for an extended period of time, the nickel-cadmium cell is in a situation where creep leakage may occur. Under these conditions there is a very small but continuing growth of an electrolyte film on the sealing surface of the positive terminal. This creep leakage, an electrochemical phenomenon, may ulti-

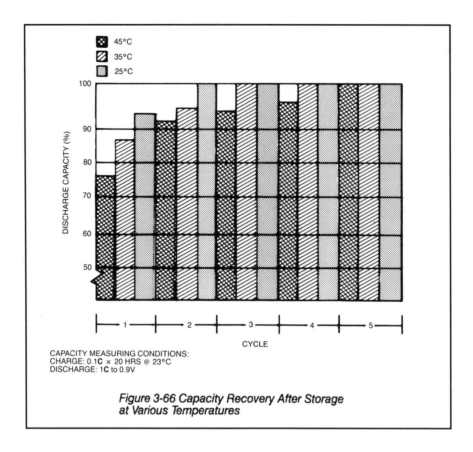

CAPACITY MEASURING CONDITIONS:
CHARGE: 0.1**C** × 20 HRS @ 23°C
DISCHARGE: 1**C** to 0.9V

*Figure 3-66 Capacity Recovery After Storage at Various Temperatures*

mately cause small amounts of electrolyte to seep around or through the vent seal. When this occurs small amounts of white crystals (potassium carbonate) will eventually appear on the outside surfaces of the positive terminal. This white material is minute particles of the potassium hydroxide electrolyte which have reacted with carbon dioxide in the air to form potassium carbonate. In rare cases, the potassium hydroxide (KOH) leakage may be large enough so that other components of the device may be exposed to potential corrosion damage from electrolyte.

This effect may also occur in circuits containing electronic battery cutoff relays in which the sensing circuit remains energized by the battery even after the main load is switched out.

Some specialized lines of nickel-cadmium cells are specifically designed to resist leakage, but storing devices with the battery in the open circuit condition is recommended to eliminate any possibility of loaded storage. Even these specialized cells are best suited to handle loaded storage early in their life since repeated cycling reduces their leakage resistance.

### 3.3.3   SUMMARY

Sealed nickel-cadmium batteries may be stored for extended periods of time. They will gradually self-discharge over a period of three to six months from full capacity at room temperature. They can continue to be stored in the discharged state without problem if they are kept in moderate temperatures (room temperature or less) and if they are stored in the open-circuit condition. *Loaded storage for extended periods is not recommended.*

# 3.4   Battery Life

Nickel-cadmium cells and batteries are suitable for a wide range of consumer, industrial, and military applications. When used properly, they provide a long-lasting, trouble-free source of rechargeable electrical energy that, in most applications, operates for many years and/or hundreds of cycles. The life of sealed nickel-cadmium batteries is influenced by actual use, the temperature, the charging and discharging parameters, and by the cell and battery design and construction. The end of useful life is heavily dependent on the requirements of the application and upon the habits and expectations of the individual user. A capacity of less than 50 per cent of the new capacity at the operating conditions is often used as an indication of the end of life for sealed nickel-cadmium cells.

Battery usage is often classified as either float or cyclic. As discussed in Section 2, float applications are those, like standby power, where the battery spends most of its time on charge waiting for use. Cyclic applications are those where the battery is used (discharged and recharged) regularly. Battery behavior, especially battery aging mechanisms, varies greatly depending on whether the battery is used in a float or cyclic application. In fact, the life of sealed nickel-cadmium cells may be measured differently depending on the end use. For float applications, calendar time is the critical parameter since the battery may be discharged for very few cycles. In cyclic applications, the passage of time usually has less to do with the way the battery ages than the number of cycles to which it has been subjected.

Some aging mechanisms are time dependent while others are cycle dependent. As the mechanisms are discussed, their relevance to either float or cyclic applications will be identified.

Life characteristics of nickel-cadmium batteries, like those of many electronic components, can be described by a bathtub shaped curve of failure rate vs. time as shown in Figure 3-67; this characteristic curve is called a *hazard function*.

The first of three distinct time periods is Zone A in Figure 3-67. It is usually identified as the infantile failure period. The typical profile shows a small and decreasing failure rate as the early failures are weeded out. Failures in this region may result from battery misuse and abuse related to the application. Quality control procedures, applied during cell manufacture, minimize the likelihood of a cell failure due to a manufacturing condition. The second phase, Zone B in Figure 3-67, is the random failures period and typically shows a very low failure rate. The third stage, Zone C in Figure 3-67, is identified as the wear-out period. During the wear-out period the battery failure rate increases with time. The term wear-out is used to describe the condition when the cumulative gradual aging effects, through oxidation or other deterioration of materials, result in eventual loss of the cell operating capability.

Battery life in any application or use, whether quantified in terms of service years or number of charge/discharge cycles, is dependent upon a combination of variables, including:

- Cell design and construction
- Charge/overcharge conditions
- Discharge conditions
- Temperature of the application
- Number of cells in the battery

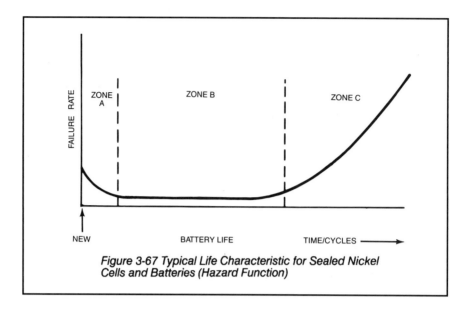

*Figure 3-67 Typical Life Characteristic for Sealed Nickel Cells and Batteries (Hazard Function)*

The realistic end of the service life of a battery occurs when the user feels that the device powered by the battery no longer meets his or her minimum acceptable standard of operation. Therefore, the end of life for a battery in any specific application depends on the expectations of the user and on the changing performance of the device itself with wear and age.

The nickel-cadmium cell can be represented by its equivalent electrical circuit as was discussed in Section 3.1. The equivalent circuit showing voltage source ($E_o$) and the series resistance ($R_e$) is shown again in Figure 3-68. In this diagram an additional parallel resistance ($R_p$) is added to schematically represent the resistance between the positive and negative electrodes, which plays a role in describing one of the failure mechanisms.

The useful life of a nickel-cadmium cell ends when either $R_e$ increases significantly or when $R_p$ becomes relatively small. These variations in $R_e$ and $R_p$ are due to

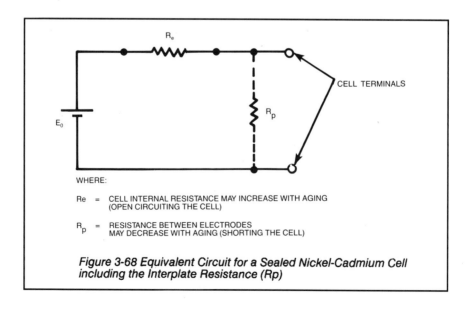

WHERE:

$R_e$ = CELL INTERNAL RESISTANCE MAY INCREASE WITH AGING
(OPEN CIRCUITING THE CELL)

$R_p$ = RESISTANCE BETWEEN ELECTRODES
MAY DECREASE WITH AGING (SHORTING THE CELL)

*Figure 3-68 Equivalent Circuit for a Sealed Nickel-Cadmium Cell including the Interplate Resistance (Rp)*

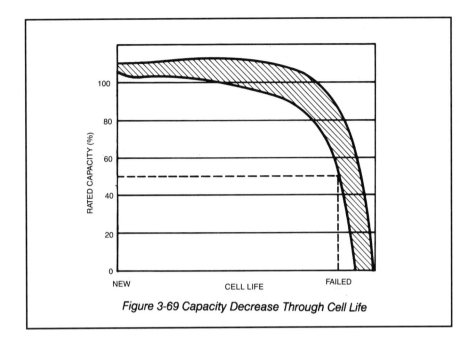

*Figure 3-69 Capacity Decrease Through Cell Life*

gradual deterioration of the cell's internal components (e.g. the separator, electrolyte, and electrodes) which change throughout their life. $R_p$ approaches zero as the cell becomes internally shorted and can no longer store energy. As $R_e$ increases significantly, the cell can no longer be effectively discharged and ceases to function. The discussion in the next section will detail the various circumstances under which $R_e$ increases or $R_p$ decreases to the degree that the cell is no longer useful.

## 3.4.1   WEAR-OUT MECHANISMS

The sealed nickel-cadmium cell typically wears out very slowly giving it a very long useful life with little loss of performance as shown in Figure 3-69. But, as the cell ages, a number of wear-out phenomena (discussed in the following paragraphs) occur, affecting performance. These include:

- Deterioration of the electrolyte
- Deterioration of the separator
- Deterioration of seal or vent integrity

### 3.4.1.1   Electrolyte Deterioration

The level of potassium carbonate in the cell relates directly to the cell's performance. Increasing carbonate concentration changes the characteristics of the electrolyte, decreasing the oxygen recombination capability of the negative electrode, and reducing the high-rate charge/discharge capability of the cell. The potassium hydroxide electrolyte used in a typical modern sealed nickel-cadmium cell initially contains a very low level of carbonate. However, organic components in the cell like the separator, seals, and vents are slowly oxidized in the presence of the electrolyte and oxygen, forming small amounts of carbonate. The carbonate accumulates as the cell ages until it ultimately reduces cell performance to an unacceptable level. Loss of electrolyte through cell venting caused by excessive overcharge or reversal can also lead to cell failure.

### 3.4.1.2   **Separator Deterioration**

Sealed nickel-cadmium cells contain a porous synthetic-fabric separator. This material provides electrical insulation and mechanical isolation between the positive and negative electrodes. A primary cause for electrical shorts in the cell (a reduction of $R_p$ in Figure 3-68) is degradation of this separator material. Degradation is cumulative. When the degradation has caused sufficient deterioration of the separator material, the separator loses its ability to insulate the electrodes. A conductive path forms and the cell shorts internally. The rate of separator degradation is a function of the integral of cell temperature over time.

A shorted cell may not always appear to have an electrical short. With the short, $R_p$ may be sufficiently high to allow most of the charge current to flow to the electrodes and charge the active material, while remaining low enough to self-discharge the cell rather quickly. Such a cell may operate acceptably if discharged immediately after charging but perform poorly on applications requiring long periods at rest (open circuit) after charging. In effect it will exhibit poor capacity retention.

When the conductive path through the deteriorated separator has a low resistance, the charge current is partially shunted and the cell cannot be fully charged. When removed from the charger, the cell will quickly discharge through the internal short-circuit, resulting in no usable capacity. The cell voltage determines the current through the internal current path (short). If the cell voltage can be increased enough by high charge currents, the shunting current is sometimes sufficient to burn off or clear the shorted contact, providing a brief extension of life (sometimes colloquially referred to as *zapping* cells).

In some applications with multicell batteries, the device may still provide satisfactory performance, even with one or more shorted cells. In other applications a single shorted cell can degrade the voltage performance of the device to an unacceptable level, at which time the battery has reached the end of its useful life. Remember the discussion regarding charge current regulation in Section 3.2.7.1. Even though the shorted battery may perform acceptably on discharge, it may be receiving much higher than expected charge currents if the charger voltage is set too close to the unshorted battery voltage.

### 3.4.1.3   **Deterioration of Seal or Vent Integrity**

The electrolyte and the oxygen gas generated during overcharge are normally contained within the cell. A typical example of a sealed nickel-cadmium cells is designed with two sealing points. One sealing component is a plastic seal ring. This seal ring electrically isolates the cover (the positive terminal) from the can (the negative terminal). The seal is made during assembly and completes the cell closure between the can and cover. The second seal point is the safety vent. Typically this is a steel disk coated with an elastomeric material which is held with a fixed force over an orifice in the cover. Neither the plastic seal ring nor the vent seal is totally immune to attack and deterioration when exposed to the combination of electrolyte and oxygen. The deterioration rate of the sealing materials is a function of temperature and of the materials used in the particular cell design.

As the seal ring and the safety vent seal (shown in Figure 2-5) slowly deteriorate, there comes a point when they no longer provide an effective gas seal resulting in hydrogen and oxygen escaping from the cell. Since the gas comes from electrolysis of the water in the electrolyte, seal leakage can eventually result in cell failure through loss of electrolyte. This failure is not usually observed because other failure modes typically occur first. As discussed in Section 3.3.3.2, loaded storage can significantly accelerate electrolyte leakage to the point where it may accelerate cell failure.

## 3.4.2   USE FACTORS AFFECTING BATTERY LIFE

The preceding paragraphs discussed the wear-out mechanisms that limit the useful life of the battery. These events basically occur whether the cell is used or not. The other key influences on battery life involve how the battery is used. When the battery is misapplied or misused it can fail much earlier than expected.

### 3.4.2.1   Effect of Improper Charging

Improper charging is the principal form of battery misuse. Any charging which results in cell venting may in turn cause an increase in the cell's $R_e$ due to water lost from the electrolyte.

An acceptable continuous charge rate (overcharge rate) allows the cell to generate only oxygen in overcharge (no hydrogen), and that oxygen is generated only in quantities that can be recombined so that the build-up of pressure does not exceed the setting of the safety vent. If the cell temperature is low (such as in a garage in winter), the ability of the cell to effectively recombine oxygen generated in overcharge is lessened. If the charge rate exceeds the recommended maximum at the cell temperature, the oxygen pressure when starting overcharge may exceed the vent setting and the cell will vent. A moderate amount of lost oxygen can be tolerated. However, if such venting continues, or is repeated, the amount of water in the electrolyte may decrease to the point that the cell will not function properly. This is sometimes referred to as dry-out. Under dry-out conditions, $R_e$ increases to the degree that the cell can no longer be effectively charged or discharged (the discharge voltage is low and the charge voltage is high).

### 3.4.2.2   Effect of Elevated Temperature

The most important environmental factor affecting normal cell wear-out is temperature. As mentioned previously, elevated temperatures cause the separator and seal materials to degrade. As this rate of degradation is a direct function of the cell temperature (at any given time), applications exposing batteries to high ambient or overcharge temperatures will experience a reduction of life from that expected at 23°C. As a rule of thumb, the cell life is reduced by one half for every 10°C increase in temperature.

Since the total amount of component deterioration is a function of the time-temperature integral, occasional brief exposures of the cell to very high temperatures of less than 100°C can be tolerated with little effect.

### 3.4.2.3   Effect of Continuous Overcharge

Sealed nickel-cadmium batteries are in widespread use in standby power or other float applications such as emergency lighting, alarm systems, computers and always ready-to-serve cordless applications.

The key factor limiting life in float applications is the cell temperature. In most standby duty equipment, the sealed nickel-cadmium battery is maintained in a fully charged, ready-to-serve state by applying a continuous charge to the battery. The temperature in overcharge, discussed in Section 3.2.4.2, can be significantly above the ambient temperature due conversion of the charging energy to thermal energy. Of course, battery temperature problems can be aggravated by location of the battery in a closed compartment or one containing other heat-generating equipment. The temperature of the cell and its internal components determines the life of the battery in these standby duty applications. Some of the degradation in life caused by high overcharge temperatures can be reduced through special cell designs intended for high-

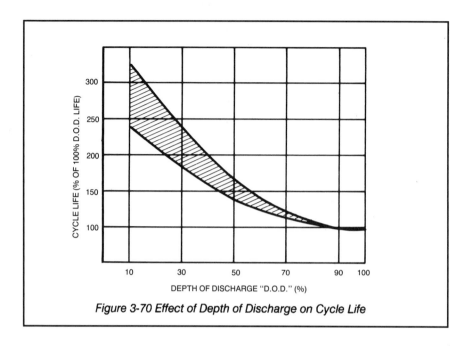

*Figure 3-70 Effect of Depth of Discharge on Cycle Life*

temperature applications. Section 3.4.3 examines the typical relationship of life versus temperature for two cell construction types.

### 3.4.2.4 Effect of Depth of Discharge on Cycle Life

Although many nickel-cadmium cells have inherent deep discharge protection, the cycle life of sealed nickel-cadmium cells is affected by the depth of discharge as shown in Figure 3-70. When a cell is fully discharged in every cycle the active materials in the electrodes are worked more, causing slight degradation to occur earlier than it would with shallow discharges. The band shown in Figure 3-70 indicates a rather broad range of expected results, particularly in the shallow depth-of-discharge region.

### 3.4.2.5 Effects of Mechanical Shock and Vibration

Nickel-cadmium cells have a rugged construction and can tolerate the shock and vibration of normal use. For example, some cells have been subjected to long-term exposure to 10g excitation in the axial and transverse directions over a frequency range of 100 to 500 Hz with no loss in performance. However, severe shock and vibration can cause damage to both battery components and cells. The result may be sudden failure due to electrical shorts within the cells or failures of the intercell connections within the battery. Under such abusive conditions the life of a battery can be abruptly terminated at any time.

### 3.4.3 EFFECT OF CELL CONSTRUCTION ON BATTERY LIFE

The life of a sealed nickel-cadmium cell depends in large measure upon its design and construction. The electrode design and method of manufacture, the choice of separator material, and the cell roll geometry have the most significant effect upon the capacity of the cell during its operating life. A typical representation of cell cycle life is shown in Figure 3-71. Cell Type A and Cell Type B indicate that different capacity trends can be expected from different types of electrode construction as a function of the number of cycles. The cells shown in Figure 3-71 were cycled

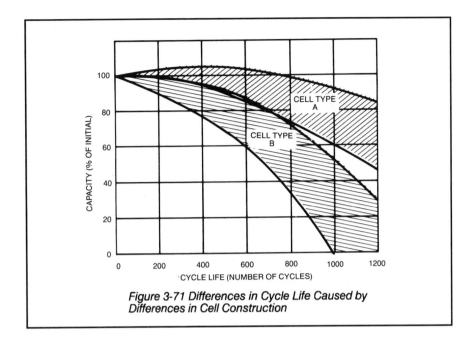

*Figure 3-71 Differences in Cycle Life Caused by Differences in Cell Construction*

through deep discharges at an average temperature of 23°C. The Type A and Type B curves in Figure 3-71 do not represent specific designs, but were chosen to illustrate that different designs and process techniques may affect cell capacity during life.

The two types of cell construction in Figure 3-72 show the calendar (non-cycle) life difference due to electrode construction, separator, electrolyte, and other design factors versus temperature. Cell Type D is able to handle high temperature exposure with limited reduction in lifetime while Cell Type E is designed primarily for applications at nominal temperature and below. Again, Type D and Type E

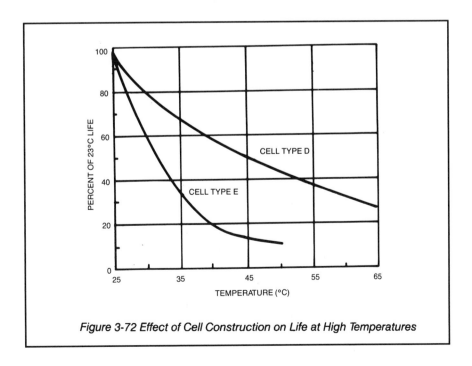

*Figure 3-72 Effect of Cell Construction on Life at High Temperatures*

curves in Figure 3-72 are not specific designs but are shown to highlight the potential differences between cell designs.

## 3.4.4    SUMMARY

Sealed nickel-cadmium batteries are a reliable long-life power source. Selection of the proper type of cell for the application and careful design to reduce the factors that stress the cell will ensure that the maximum life is obtained from the cell or battery.

# 3.5   Application Information

Sealed nickel-cadmium batteries and cells are used in a variety of products ranging from portable appliances to computer power backup. The key to their selection for such a diversity of applications is the unmatched combination of performance, flexibility, and ruggedness they provide. Earlier sections provided details on the attributes of the sealed nickel-cadmium cell. This section focuses on some of application benefits resulting from these attributes. It also presents a quick summary of typical applications for sealed nickel-cadmium cells.

## 3.5.1   FEATURES AND BENEFITS OF SEALED CELLS

In reviewing the features of sealed nickel-cadmium cells and batteries presented in earlier sections, it is clear that these batteries are particularly well suited to applications where the convenience of a self-contained, always ready power source increases the value of the end-product. In rechargeable battery applications needing high power delivery or high power density, a nickel-cadmium battery is usually the only logical choice.

Sealed nickel-cadmium batteries are used in diverse applications due to their unique characteristics and capabilities. Among the significant features that lead to their position as the preferred power source are:

- High energy density
- High-rate discharge capability
- Fast recharge
- Long operating life
- Long storage life
- Rugged construction
- Operation over a broad range of temperatures
- Operation in a wide range of environments
- Operation in any orientation
- Maintenance-free use
- Continuous overcharge capability
- Consistent discharge voltage

An understanding and appreciation of the product benefits derived from this unique combination of product features is important. These special capabilities open the way to the design and manufacture of a host of products that are practical, convenient, or economical only with the sealed nickel-cadmium battery system.

### 3.5.1.1   High Energy Density

The watt-hours per kilogram (or per cubic centimeter) provided by the nickel-cadmium battery are the highest available for a widely available rechargeable system. A lead-acid battery capable of delivering the same amount of energy weighs considerably more and requires more space than a nickel-cadmium battery. Thus, specifying sealed nickel-cadmium batteries allows the design of small, lightweight devices such as instruments, video cameras and shavers.

### 3.5.1.2 High-Rate Discharge Capability

Sealed nickel-cadmium batteries can deliver energy at very rapid rates. High-rate discharge capability makes the nickel-cadmium battery ideal for use in high-power devices such as power tools. In these applications a high-rate discharge capability leads to small, light, and economical battery-powered products. And these cells are rugged enough to withstand repeated cycles with very high discharge rates. More information on discharge capabilities is presented in Section 3.1.

### 3.5.1.3 Fast-Recharge Capability

One key to the success of cordless products is the ability of current nickel-cadmium batteries to recharge quickly. Most rechargeable nickel-cadmium battery applications have traditionally used a relatively low-charge rate which may require 16 to 20 hours to achieve a full charge. Now specially designed sealed nickel-cadmium batteries that can be charged much more rapidly are available.

Quick-charge batteries can be completely recharged at the 0.33C rate in four to five hours, generally without the need of charge controls. They can also be over-charged at up to the 0.33C rate, further simplifying charger design.

Fast-charge cells allow the user to recharge the battery in very short periods of time, normally within an hour. With their fast-charge capability, the use of cordless electric products is possible without advance planning or preparation. Fast charging also permits the use of the full capability of the battery several times a day, reducing the need for spares or larger batteries. Fast charging is discussed in detail in Section 3.2.

### 3.5.1.4 Long Operating Life

Nickel-cadmium batteries have a very long operating life, as measured either by number of charge/discharge cycles or by years of useful life. Whether the battery is actively used by repetitively charging and discharging, or is maintained on charge in a ready-to-serve condition, nickel-cadmium cells offer a long, trouble-free life. Sealed nickel-cadmium batteries will normally provide hundreds of charge/discharge cycles or operate for many years in a ready-to-serve standby function. A detailed discussion of battery life is presented in Section 3.4. The long cycle life makes nickel-cadmium batteries an excellent value for high-frequency use.

### 3.5.1.5 Long Storage Life

The sealed nickel-cadmium cell can be stored for extended periods of time at room temperatures in either a charged or an uncharged condition with virtually no degradation in capability. Even after long-term storage, just one or two normal charge/discharge cycles returns the battery to full capacity. This makes these cells especially practical in applications where the battery is used only sporadically while spending long periods of time at rest in the discharged state. Battery life in storage is similar to that in use although with a lower rate of degradation. Storage is discussed in Section 3.3.

### 3.5.1.6 Rugged Construction

The sealed nickel-cadmium battery is a very rugged device, both physically and electrochemically. The cells possess good resistance to shock and vibration. The nickel-cadmium battery lends itself to applications in which it may be subjected to temperature extremes, tough physical use, or other demanding requirements.

### 3.5.1.7 Operation Over a Broad Range of Temperatures

Nickel-cadmium batteries have excellent performance characteristics over a wide range of operating temperatures. They deliver usable capacity in applications where temperatures may drop to -40°C or rise to 70°C. Cells specially designed for high-temperature applications are usually recommended for applications in which the battery consistently remains above 50°C.

### 3.5.1.8   Operation in a Wide Range of Environments

Sealed nickel-cadmium cells operate in environments with a relative humidity range normally encountered in commercial applications without noticeable effects on performance or degradation of life. Sealed nickel-cadmium cells also operate equally well in a full vacuum or in a positive pressure environment. They have even powered drills in space and on the moon!

### 3.5.1.9   Operation in Any Orientation

Sealed cells can be mounted and operated in any position or attitude. The combination of sealed construction, absorbent separator, and highly porous electrodes holds the electrolyte so that normal ionic conduction required in the charging and discharging reactions occurs regardless of the cell orientation. This feature permits product design flexibility for portability.

### 3.5.1.10   Maintenance-Free Use

Nickel-cadmium sealed-cell batteries are a good choice for applications in which periodic maintenance would be difficult or costly. Since they require no maintenance, sealed cells are often permanently wired into a product. The gases generated during normal operation are recombined within the cell. Sealed cells perform repeatedly through normal duty cycles with no loss of active material or electrolyte.

The sealed cell is an install-and-forget power source. With the exception of periodic charging, it will perform virtually without attention throughout its life.

### 3.5.1.11   Continuous Overcharge Capability

The nickel-cadmium sealed-cell battery is designed to accommodate extended charging at recommended charge rates, with no noticeable effect on performance or life. Very simple and inexpensive chargers can be used when charging at standard rates that fully charge the battery in 16-20 hours. Where a faster charge is desired, special quick-charge cell designs are available which will tolerate continuous overcharge up to 0.33C rate and can be recharged in four to five hours with a similarly inexpensive charger. The ability to accept continuous overcharging using either the standard or quick-charge rates is valuable for those applications where the product must be ready to operate upon demand. Because of their ability to handle continuous overcharge, sealed nickel-cadmium batteries are widely used in standby power applications to automatically provide temporary service during power outages. Examples are emergency lighting, alarms, and microprocessor memory holdup applications.

### 3.5.1.12   Consistent Discharge Voltage

Sealed nickel-cadmium batteries have a nearly flat discharge voltage profile throughout the major portion of the discharge. This is useful in applications that require a high voltage to operate efficiently, such as a DC motor-driven product, or where little variation in electrical power output is important. Examples of such applications include cordless drills, tape players, portable televisions and shavers.

## 3.5.2   ECONOMIC CONSIDERATIONS

Cost is often a major factor in selecting a battery system. Cost concerns may include the battery purchase price, the cost of associated systems such as the charger, and the cost of using the battery system throughout its life. In addition to the direct costs associated with the battery, a major potential economic issue for standby systems is the cost of alternatives or of loss of power if the battery is not used. Depending on the end product, the first cost of the system may be the principal economic interest or the total life-cycle cost may be the dominant concern. In either case the designer can best create a cost-effective battery system with a working knowledge of the cost of nickel-cadmium batteries including the effects of duty cycle, use pattern, and application variation on the system's economics. The following discussion reviews some of the economic considerations in using sealed nickel-cadmium batteries.

### 3.5.2.1   Battery Duty Cycle

In previous sections the various discharge characteristics and recharging rates which are part of a nickel-cadmium cell's normal duty cycle were discussed in detail. Three rates of charge were defined:

(a) Standard-charge — The rate which requires overnight recharging time;

(b) Quick-charge — a moderately rapid rate which recharges cells in several hours; and

(c) Fast-charge — which uses the high-current capability of nickel-cadmium sealed cells by recharging them in about one hour or less.

For a rechargeable battery system, the duty cycle or use pattern will have a significant impact on the required battery rating, hence size and cost. In Figure 3-73, three different duty cycles are shown. Each duty cycle requires the same amount of energy from the battery in a 24-hour period.

The constant energy required in Cycle A in Figure 3-73 for an eight-hour period suggests a battery capacity equal to the total capacity required, say 2.0 ampere-hours, for the eight-hour period (250mA x 8 hours = 2.0 ampere-hours). This means that the battery capacity required to satisfy the requirements of Cycle A in Figure 3-73 needs to be at least 2.0 ampere-hours at an eight-hour drain rate. The battery would be recharged during the 16-hour off period at the 0.1C rate (200mA) so that it would be fully charged for the next discharge period. This duty cycle is an example of a standard-charge rate application.

The energy required for Cycle B of Figure 3-73 is also 2.0 ampere-hours over the 24-hour period, but it is split into two four-hour periods of 1.0 ampere-hour each. The drain rate of 250mA is the same. In this case, a smaller battery with a quick recharge capability could be chosen. A 1.0 ampere-hour battery with a 0.3C ampere recharge rate capability of 300mA will suffice. This significantly reduces the battery size, weight, and cost. The battery, after a four-hour discharge, would be recharged during the next four hours and be ready for the next discharge. This is a quick-charge cell application.

In Cycle C of Figure 3-73, the same energy and drain rate (2.0 ampere-hours at 250mA) is delivered to the load, but with at least one hour off between each one-hour discharge period. A battery and charger system capable of one-hour recharge reduces the battery ampere-hour rating required. The battery must deliver 250mA for one hour every other hour. A 250mAh battery (dramatically decreasing the size of the battery), along with a charger that is able to terminate the fast-charge current of 250mA (1C) once the battery is charged, could meet this need. The charger cost is likely to be higher than for quick or standard-rate charging. This may offset some or all of the lower battery cost.

As these three examples show, taking advantage of a short duty cycle by using quick or fast-charge cells can reduce battery cost, size, and weight. The charger cost, however, increases with higher power requirements and with greater charge-control

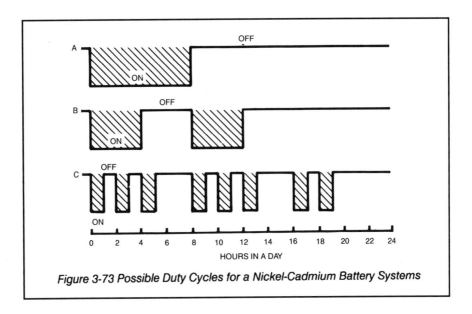

*Figure 3-73 Possible Duty Cycles for a Nickel-Cadmium Battery Systems*

requirements. These trade-offs should be evaluated in selecting both cell and charger system.

### 3.5.2.2 Voltage vs. Watt-Hour

The battery cost can often be minimized by optimizing the voltage-versus-watt-hour relationship. Many applications simply need a certain energy (voltage × amp-hours) from the battery. In these cases, using a small number (lower voltage) of large-capacity cells is an alternative to using a greater number (higher voltage) of smaller capacity cells. Since the cost per amp-hour is usually less in larger cells and there are fewer cell interconnections required, the large-cell battery will usually provide the same watt-hours at a lower cost than the higher voltage, small-cell battery. The larger capacity cells may also offer higher reliability by having fewer cells per battery.

### 3.5.2.3 Life-Cycle vs. First Cost

Selection of batteries and charger systems often depends on whether the design goal is simply to minimize first cost or to obtain the lowest total cost over the duration of the application. If the battery's initial cost is the only consideration, the nickel-cadmium battery may not be the cheapest battery system. However, evaluated over the operating life of the product to be powered, it often becomes the most economical choice by virtue of its longer life. Many applications only need the original set of nickel-cadmium batteries while other battery types would need to be replaced one or more times. Other costs, including additional maintenance expenses, costs for special mounting and storage systems, and costs for loss of function due to battery failure, are typically incurred by other battery systems, but not by nickel-cadmium batteries. The result is the sealed nickel-cadmium battery normally providing a lower life-cycle cost than other battery systems.

#### 3.5.2.3.1 Cost Per Watt-Hour

The first cost (or replacement cost) of a battery is normally evaluated in terms of cost per watt-hour.

The nickel-cadmium cell is rated 1.2 volts DC. Hence, 1.2 volts times the ampere-hour capacity of the cell gives the watt-hour rating of the cell. This product, multiplied by the number of cells in the battery, yields the battery watt-hour rating as shown in the equation below. The cost of the battery can then be reduced to a common denominator of cost per watt-hour:

$$\text{cost/watt-hour} = \text{cost}/(1.2 \times N \times C)$$

Where:    1.2 = cell voltage
              n = number of cells
              C = rated capacity of each cell (amp-hours)

### 3.5.2.3.2 *Cost Per Watt-Hour Per Cycle or Per Year*

As stated earlier, the true cost of a rechargeable battery is the cost over the life of the battery. Sealed nickel-cadmium batteries are designed so that in most applications they will deliver hundreds of charge/discharge cycles over many years. This long operating life must be evaluated when determining the true lifetime cost of the system. A truly accurate comparison of costs between two competing battery systems will consider all of the costs occurring during the application's life for each system. A method such as discounted cash flow analysis may be used to account for the differences in timing between the two sets of expenses. The end result is a cost difference for selecting one battery system over another.

Even without the detailed analysis described above, simply considering the battery cost per year or per cycle often gives an excellent indication of the life-cycle cost.

## 3.5.3   ELECTRICAL CONSIDERATIONS

The previous sections have described the electrical implications of the discharge and charge processes. There are, however, some additional electrical considerations relating to the application of sealed nickel-cadmium cells and batteries. These are discussed in the following subsections.

### 3.5.3.1   Cell Voltage—General Overview

The general cell-voltage response during a typical cycle is shown in Figure 3-74. In this representation a fully discharged cell is charged at the 0.1C rate in a 23°C ambient into the overcharge state, rested, discharged, and then rested again.

Figure 3-74 shows the voltages for each phase of this typical cycle. The voltage of a nickel-cadmium cell can vary from about 0.6 volts per cell up to about 1.5 volts depending upon the cell's state of charge and whether it is being discharged, charged, or is at rest. The useful discharge voltage range is normally 1.3 volts to 0.9 volts in typical applications. The normal standard (0.1C) rate charging voltage range is from 1.3 to 1.5 volts per cell. Occasional deviations from these limits may be observed in extreme temperature environments, at high charge or discharge rates, when the cell is new, or when the cell has been stored for an extended period. (See Section 3.2 on charging.)

The discharge characteristics of a nickel-cadmium battery are quite different from the ordinary dry-cell battery. Figure 3-75 compares the discharge voltage characteristics of carbon-zinc and alkaline primary cells with the rechargeable nickel-cadmium cell. The comparison shown is for a discharge current of 800mA where all cells are "D" size. The voltage of the carbon-zinc cell falls quite rapidly as it is discharged, while the nickel-cadmium cell discharge voltage is nearly constant for most of the discharge. Even ignoring the voltage drop, the carbon-zinc cell lasts only about half as long as the nickel-cadmium cell. The alkaline cell is a far better performer than the carbon-zinc cell. The voltage drop during discharge, although still significant, is not as dramatic as the carbon-zinc cell. The battery capacity for an

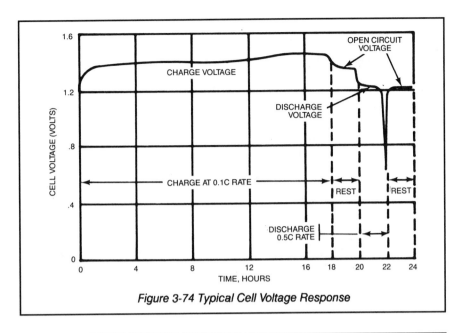

Figure 3-74 *Typical Cell Voltage Response*

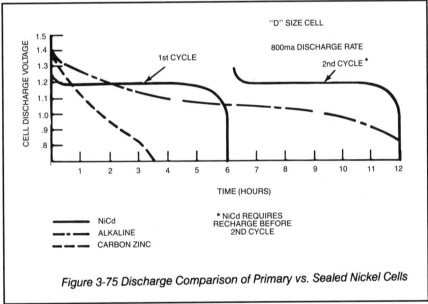

Figure 3-75 *Discharge Comparison of Primary vs. Sealed Nickel Cells*

alkaline cell is about twice the capacity that can be obtained from a single discharge of a nickel-cadmium cell. Of course the nickel-cadmium cell can be recharged while the alkaline cannot. (See Section 3.1 for further data on nickel-cadmium cell discharge performance.)

### 3.5.3.2 Designing for the High-Rate Capability of Nickel-Cadmium Batteries

Since nickel-cadmium batteries are capable of very high rate discharge, products using them must be designed so that neither the battery nor the conductors will overheat. A short circuit across a fully charged nickel-cadmium battery can result in currents up to 100C. The resistive heating from such a large current can cause temperatures to rise quickly with the possibility of a safety hazard.

Good design practice uses circuitry and physical barriers to minimize the possibility of short circuits while also providing protection (such as fusing) that minimizes the consequences should a short circuit occur.

### 3.5.3.3    Low State of Charge Indicator

As discussed in detail in Section 3.1, the voltage and internal resistance of nickel-cadmium batteries varies only slightly with state of discharge so it is difficult to use these parameters to determine state of charge. It is possible to determine when a nickel-cadmium battery is fully charged or when it is fully discharged. Between these extremes, variation of the battery voltage and internal resistance with temperature, battery history and other environmental factors usually make them poor indicators of state of charge.

### 3.5.3.4    Proper Voltage Selection

Nickel-cadmium batteries can, in certain applications, fail to meet performance expectations due to improper selection of the number of cells in the battery pack. In certain use patterns and application environments the battery discharge voltage may be depressed. This is caused by previous history, temperature, rest time, and time on charge. (See Section 3.1.5.2) However, these voltage variations may be accounted for during product design.

In the case of a 5-cell battery, for example, the battery is fully discharged at the 1C rate when its voltage falls to 4.5 volts (0.9 volts per cell or 0.75 MPV). If the minimum operational voltage for the end product is 5.4 volts, the battery may fail to deliver full capacity in some circumstances (such as low temperature), since the 5.4 volt cutoff represents 1.08 volts per cell, or 87% of MPV. A 6-cell battery is needed in this example to achieve full capacity utilization under all environmental conditions. For it, the minimum required voltage (RAV) of 5.4 volts is 0.9 volts per cell which means that the battery will provide its entire useful capacity at voltages above the RAV.

Proper design (RAV ≤ 0.75 × MPV) also avoids possible voltage depression (sometimes called memory) without the need for conditioning cycles.

In general, the higher the drain rate the lower the cutoff voltage that should be used. The formula provided in Section 3.1.5 estimates the cutoff voltage which permits full utilization of battery capacity.

### 3.5.3.5    Maximum Power Discharge

Most battery applications focus on getting the full capacity out of the battery at some relatively low discharge rate. These are energy-type discharges where the goal is to maximize the energy delivered by the battery. However some applications, notably engine starting, are more interested in obtaining high instantaneous rates of power transfer. These power-type applications need to be treated somewhat differently from the energy-type of discharges that are generally the subject of this Handbook.

When discharged at relatively low rates (2C and less), the vast majority of the electrical energy being generated in the cell is delivered to the external circuit. These low rates are typical of energy-type discharges. However, as the discharge rate is increased, more and more voltage is dissipated by the effective internal resistance of the cell, i.e. an increasing fraction of the energy goes into internal heating of the cell. A smaller and smaller portion of the converted energy, therefore, is delivered to the external circuit. The actual power delivered to the external circuit, however, increases until it maximizes then decays back to zero at the short circuit current.

The effect of discharge rate on power delivery is shown in Figure 3-76. This

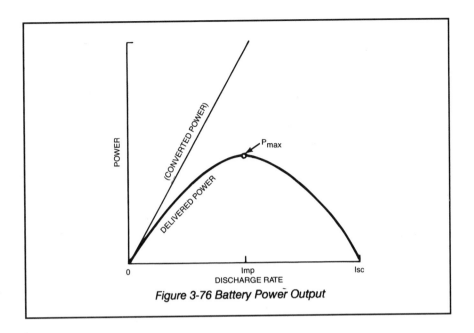

Figure 3-76 Battery Power Output

curve illustrates that for maximum power delivery from a given size cell, the discharge rate should be approximately one-half the short-circuit current. This is the conventional matched-impedance operating point, where load resistance equals source resistance, for maximum power transfer.

At the maximum-power operating point ($P_{max}$) approximately one-half of the converted power is actually delivered to the external circuit, while the other half is dissipated internally in the cell as heat. The $P_{max}$ point occurs at a maximum power current ($I_{mp}$) discharge rate. $I_{mp}$ for a source with a linear load regulation line is one-half the short-circuit current ($I_{sc}$) and coincides with a delivered voltage which is half of $E_o$, the effective cell no-load voltage. Operating the cell at the $I_{mp}$ rate, therefore, produces the maximum deliverable power and an equivalent amount of internal heating. The delivered capacity at this rate is far less than at the battery's **C** rate because half the energy is going to heat the cell. Operating at $I_{mp}$ does, however, permit using the minimum size of cell for a particular power requirement.

Prolonged or repeated operation at maximum power rates requires caution. Since approximately half of the converted energy goes into heating the battery, the cell temperature rise may be excessive if the battery is fully discharged at the $I_{mp}$ rate or if partial discharges are repeated without allowing time for cooling between them. A total discharge at the $I_{mp}$ rate, without cooling, could raise internal temperatures by more than 50°C. For this reason, the application engineer may have to consider special heat sinks or forced air cooling to dissipate the heat.

$I_{mp}$ and $R_e$ are related by:

$$I_{mp} = E_o/(2R_e)$$

In fact, $I_{mp}$ has some advantages over $R_e$ in design calculations. First, $I_{mp}$ is a property of the basic cell which, unlike $R_e$, remains independent of the number of cells connected in series. $R_e$ for a battery is approximately equal to N times $R_e$ of the individual cell. Second, both the capacity and the weight of a cell are approximately proportional to its $I_{mp}$ value, but inversely proportional to its $R_e$ value. When cells are placed in parallel, capacity, weight and $I_{mp}$ are all the sum of the individual values, while $R_e$ is the inverse of the sum of the individual cell values. Third, the maxi-

| DISCHARGE OF CELL A AND CELL B IN PARALLEL | COMBINED EFFECT |
|---|---|
| NO LOAD VOLTAGE | SAME AS EITHER CELL |
| CAPACITY | SUM OF CAPACITIES OF CELL A AND CELL B |
| $I_{mp}$ | SUM OF $I_{mp}$ (CELL A) and $I_{mp}$ (CELL B) |
| $R_e$ | $\dfrac{R_e \text{ (cell A)} \times R_e \text{ (cell B)}}{R_e \text{ (cell A)} + R_e \text{ (cell B)}}$ |

*Table 3-4 Parallel Discharge Effects*

mum deliverable power from a cell is proportional to its $I_{mp}$ capability so that an $I_{mp}$ rating for a given cell directly indicates its maximum power capability.

### 3.5.3.6    Parallel Discharge

Nickel-cadmium sealed cells are generally connected in series to achieve a desired voltage. The capacity available for discharge is then governed by the cell size (amp-hours) selected. However, batteries may be connected in parallel during discharge to obtain greater capacity.

Connecting two nickel-cadmium batteries of equal or unequal capacities (cell size) in parallel results in the combined effect shown in Table 3-4.

Since the terminal voltage of both batteries must be equal, no instabilities exist when sharing the discharge load current.

A problem with parallel-connected batteries arises in charging. Parallel charging could easily result in improper sharing of charge current, resulting in battery damage. The best approach is to use a series/parallel switch or electronic system to allow only discharging in parallel. See Section 3.2.7.4 for a more comprehensive discussion of charging for series/parallel batteries.

### 3.5.4    PHYSICAL CONSIDERATIONS

The location and mounting of the battery may do much to determine the success of an application. The major considerations are discussed below.

### 3.5.4.1    Proper Location of the Battery with Respect to the Charger

In many applications, wiring runs are simplified if the charger and battery are mounted in close proximity. But, for most chargers, this may reduce battery life. Transformer/rectifier-type chargers generally produce a significant amount of heat. If this heat is transferred to the battery the increased battery temperature may shorten battery life. Therefore, it is usually best to locate the transformer at a distance from the battery, such as having the charger supported by the wall outlet with a cord to the battery, or to thermally insulate it from the battery.

Capacitive chargers produce less heat than equivalent transformer chargers and may be a solution if it is important to mount the charger close to the battery.

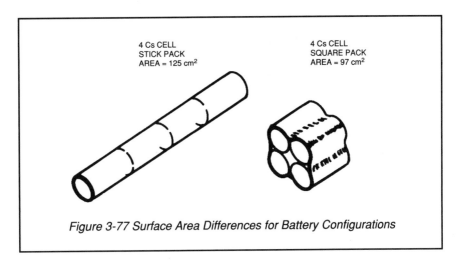

*Figure 3-77 Surface Area Differences for Battery Configurations*

Capacitive charging is discussed in Section 3.2.7.2 and a typical circuit is shown in Figure 3-59.

### 3.5.4.2   Battery Enclosures and Form Factor

Design of an enclosure or container for the battery centers around two concerns: proper selection of materials and design for adequate heat transfer.

The most common battery enclosures are made from plastic materials that are resistant to alkaline solutions and have a high impact strength. Metal housings are sometimes used, but metal requires careful design and assembly to avoid shorting of the cells in the battery pack. Aluminum is not recommended for enclosures because if cell leakage does occur, the electrolyte will react with the aluminum.

The battery enclosure may also have a significant impact upon battery life if not designed properly to prevent the cell's temperature from rising significantly in overcharge. Good battery form factor allows for air movement around cells to help reduce temperature rise in overcharge. As shown in Figure 3-77, a four-cell stick (cells connected end-to-end) has a significantly greater surface area for heat dissipation and therefore lower overcharge temperature than would four cells in a square configuration.

### 3.5.4.3   Battery Assemblies

Many consumer devices that use primary cells create batteries within the unit as the consumer adds single cells. This approach is generally not recommended with sealed nickel-cadmium cells. Complete battery assemblies in overall plastic shrink wrap, vacuum-formed or injection-molded plastic cases, with wire leads and connectors, are more suited to rechargeable devices. Since replacement is comparatively rare with nickel-cadmium batteries, more effort can be placed in developing the optimum battery design rather than worrying about providing for frequent replacements.

In rechargeable batteries, replacement of individual cells as they wear out is usually not practical as all the cells have a similar life. Also, individual cell replacement may result in mixing cells of different internal construction into one pack. For these reasons, batteries should normally be replaced as a unit.

Most manufacturers assemble a wide range of batteries that vary in geometry, packaging material, interconnection, and in battery terminations. The variety of batteries readily available in standard forms allows for design flexibility without requiring custom assembly.

#### 3.5.4.4 Cell Interconnection

The key to successful interconnections between cells is to minimize the voltage drop in the interconnection. For this reason, welded interconnections are usually preferred to the pressure contact holders common with primary (throw-away) cells.

##### 3.5.4.4.1 Welded Interconnections

Typically sealed nickel-cadmium batteries are assembled using welded interconnections. These provide a rugged, low-resistance connection between cells. Normally cell interconnects are metal strips. If the product is likely to encounter high-vibration environments, the solid strips may be replaced with either metal braid or wire interconnects. Wire intercell connections have weld tabs soldered to each end so that they can still be welded to the cells.

*Never solder directly to the nickel-cadmium cells, only to solder tabs or wire leads. A hot soldering iron placed directly on the cell is likely to cause seal ring and vent seal damage as well as damage to the separator systems.*

##### 3.5.4.4.2 Pressure Contacts

Pressure contacts are generally not as satisfactory as welded contacts because they tend to be higher resistance. However in some applications they are unavoidable. If they are used, the following are some suggestions on ways to minimize problems.

Since surface deposits interfere with the clean contact needed for minimum resistance, contact materials should be selected to be conductive, but inert. For example nickel-plated contacts tend to be less affected by oxidation than either copper or steel. Problems with surface deposits will be reduced if the installation process "wipes" the deposits off giving a clean contact. Frequently replaced primary cells regularly wipe the contacts. Unfortunately, the nickel-cadmium rechargeable battery usually gives years of service before needing replacement; thus, contact wiping through cell removal seldom occurs.

Flat-surface to flat-surface contacts often prove unsatisfactory. A much better solution is the point contact obtained by providing a dimple on one surface and a mating protrusion on the other. This provides a positive mechanical positioning and interface not possible with flat contacts.

Some form of spring-loaded contact is typically required to ensure to minimize resistance between contact and cell. Normally, a stronger spring is better so that the cell and terminals approach an "interference" fit.

Finally, pressure contacts should only be used in low-current applications where resistance losses in the contact are likely to be less significant.

### 3.5.5 ENVIRONMENTAL CONSIDERATIONS

The environment to which the cell is exposed may have major effects on the way it performs. In particular, the cell operating temperature is a critical determinant of cell performance.

#### 3.5.5.1 Temperature

High and low temperature applications require some special considerations. Incremental charge acceptance is reduced at elevated temperatures. Charging time, therefore, needs to be extended to gain maximum advantage of available capacity. However, even with extended charge time, the battery may not be capable of storing as much capacity as if charged at room temperature. (See Section 3.2.7.6 on high-temperature charging.) Two high temperature alternatives are available. One is to charge at room temperature and discharge at the high temperature. The other is to oversize the battery to allow for the reduced available capacity at the operating temperature as shown in the applicable specification sheet. In both cases, a cell designed for the high-temperature environment will extend the longest service life, which still may be less than life at room temperature. High temperature environments will also increase the rate of self-discharge which may affect storage planning.

Ventilation of the battery pack is important to minimize the adverse effects of high temperatures that may develop from the self-generated heat of a large battery when continuously overcharged. See Figure 3-31 for calculating the temperature rise of a battery in overcharge.

Low cell temperature will also reduce the actual cell capacity. Charging at low temperature requires special controls that will reduce the rate of charge as the temperature becomes lower and which will allow little or no charging below –20°C. The charge current values for low-temperature charging are described in Section 3.2.7.5. Low-temperature discharging is acceptable, but battery capacity must be derated from room-temperature capacity; therefore run time is reduced. Delivered voltage may be reduced by the increase in $R_e$, particularly at high discharge rates. Again, proper design should eliminate voltage problems during use.

At low temperatures, the discharge rate has a great effect on deliverable capacity. Both the midpoint voltage and available capacity will be significantly reduced as discharge rate is increased. Self discharge is reduced at low temperatures.

## 3.5.6   OTHER CONSIDERATIONS

Other concerns related to logistics and charger interfaces are described below.

### 3.5.6.1   Handling Batteries in the Charged State

Nickel-cadmium batteries are typically shipped in a discharged state. Charged batteries can present a handling as well as safety problem because of their ability to deliver large amounts of current that can, when shorted, cause burns. Manufacturers do provide charged batteries for special customer requirements, but this requires special precautions by the purchaser who specifies charged batteries. If an inspection process is designed around using charged batteries, inventories must be accurately controlled to avoid in-process problems resulting from self-discharge. Also, special handling procedures are mandatory for charged cells. (See Section 3.7 on safety.)

In-process charging is the common solution when charged batteries are required, allowing ease in handling and flexible inventory control. The device may be tested from console power while at the same time slightly charging the battery. In this case, battery power is then used only to verify that the battery is in the circuit and to check parameters such as correct polarity wiring by observing motor rotation.

### 3.5.6.2   Inventory

Batteries should be used on a first-in, first-out (FIFO) system. Battery life degradation is a function of time, even if the battery is never used. As temperature increases, the degradation rate of the battery increases, making it desirable to keep inventory between 0°C and 30°C when practical. Storage is covered in detail in Section 3.3.

### 3.5.6.3   Agency Listing of Chargers

Charging from a wall outlet-type electrical source usually means agency safety evaluation or listing of the charger will be required. One advantage of the remote charger is that agency listing can, for many charger types, be obtained separately from the product containing the battery. In many instances this eliminates any need for the

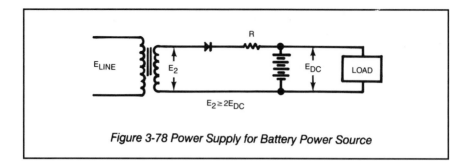

*Figure 3-78 Power Supply for Battery Power Source*

battery-powered device to be evaluated by the agency, as it is not directly connected to AC line voltage. Charger manufacturers frequently offer agency-listed standard charger types.

### 3.5.6.4   Detachable Chargers

When a detachable charger is disconnected from a battery the terminals of the battery may be exposed and could be shorted by pencil tips, paper clips, coins, etc., resulting in a rapid high-current discharge. Methods of preventing unwanted rapid discharge include: 1) placing the rectifying element (diode) within the battery rather than the charger, 2) placing a current-limiting device (fuse or resistor) between one cell within the battery and the charging terminal, and 3) design of a polarized connector which eliminates the possibility of simultaneous contact of the battery power terminals.

### 3.5.7   TYPICAL APPLICATIONS FOR SEALED NICKEL-CADMIUM BATTERIES

While applications for sealed nickel-cadmium batteries are extremely diverse, there are certain key product types that use large quantities of cells. The major applications are described below.

### 3.5.7.1   Standby Power

Batteries are the backup power source for uninterruptible power supplies used to provide emergency power for alarms, lighting systems, and computers in case of a power failure.

Nickel-cadmium batteries are ideal for these applications as they are capable of continuous overcharging and can tolerate a wide range of temperatures. Figure 3-78 shows a battery as an integral part of the power supply. The battery acts not only as an alternate or portable power source, but may also provide sufficient filtering to replace the common electrolytic filter capacitor.

### 3.5.7.2   Emergency Lighting and Alarms

Applications using nickel-cadmium batteries in room temperature environments (0°C to 35°C) need little special consideration other than the heat generated by continuous overcharge current. Systems that are mounted in exterior environments where very low temperatures occur will require charge control.

High temperature environments create special problems. Cell life is reduced by high temperatures. This must be considered when designing for battery locations near a fluorescent ballast or in an enclosure having poor heat transfer. In addition to a potential reduction in operating life at high temperatures, the charge acceptance is less. In a lighting fixture where the battery is continuously held at a high temperature (40°C to 60°C) due to lamp/ballast heat, the battery may not attain a capacity as high

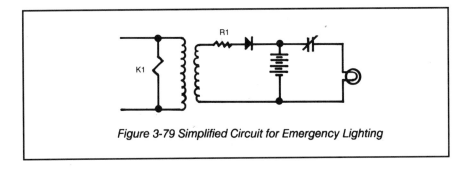

*Figure 3-79 Simplified Circuit for Emergency Lighting*

as the 23°C cell specification rating (see Figure 3-24, Section 3.2.2.1.). When charged at high temperatures the battery will deliver only a portion of its room temperature capacity. If the battery continues to receive charge current after the lamp/ballast is turned off, thereby allowing the battery temperature to drop to room temperature at some point in the day, full capacity may be achieved. The chemistry used in cell manufacture influences high temperature charging so the cell product specification for charge acceptance vs. temperature should be examined.

Figure 3-79 shows a simplified circuit for emergency lighting which provides power to the lamp when AC line power is interrupted. The system includes a charging circuit to keep the battery charged when AC line voltage is present. The relay contact (K1) is open when line power is on but closes when the relay coil is de-energized due to a power outage. When the relay contacts close, the battery supplies current to the emergency lamp. When AC power returns to the circuit, relay contact (K1) opens and the lamp is turned off. The battery is again charged, giving a simple and completely automatic emergency lighting system.

### 3.5.7.3  Electronic Loads

Batteries are frequently mounted directly on printed circuit (pc) boards as a local source of uninterruptible power for electronic memory. Some special charging circuits have been developed to serve these applications, with some examples shown in Figure 3-80.

One special precaution that should be followed for these applications is to avoid storage of the battery while it is connected to a load. Nickel-cadmium batteries should be stored in an open-circuit condition. Even the very small current drain of a microchip can have an adverse effect on the battery in storage. Once the battery is completely discharged and the battery voltage is held near zero, this loaded-storage condition can induce leakage of electrolyte from the cell. Therefore the load circuit should be designed so the battery will be open-circuited during long-term storage. Many times this is accomplished by making the battery circuit incomplete until the pc board is inserted into the device or insulating the battery with a non-conductive material that is removed when the device is placed in operation. Since this is not practical in all applications, some manufacturers have developed special cells that are highly resistive to the adverse effects of loaded storage.

An example of using the sealed nickel-cadmium battery for standby power is a small computer which uses MOS semiconductor memory. Power input to the memory must be maintained above a certain voltage to avoid loss of data. If voltage to the memory drops below a critical level, the entire contents of the memory must be retrieved and reentered once normal power is resumed. The battery provides proper memory support power. The block diagram in Figure 3-81 illustrates this application.

One method of accomplishing the instantaneous transfer to battery power is

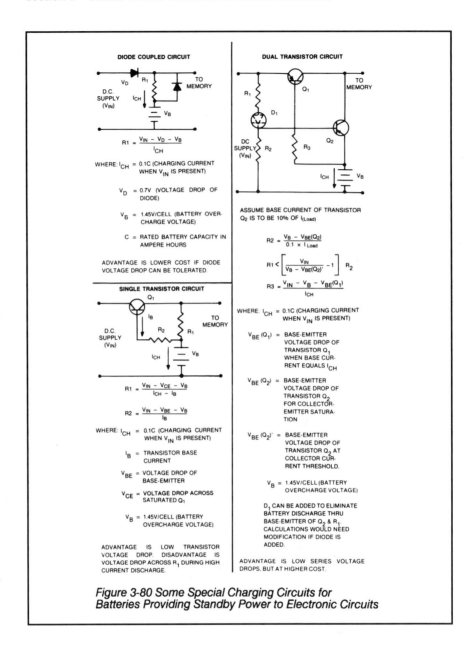

*Figure 3-80 Some Special Charging Circuits for Batteries Providing Standby Power to Electronic Circuits*

shown in Figure 3-82. The battery is selected so that its charge voltage minus the threshold voltage of the rectifier is slightly below the normal DC voltage level, $(E_2)$, of the power supply when normal AC power is on. Rectifier $(CR_1)$ isolates the battery from the equipment when the AC power is on. When the power is interrupted, $E_2$ begins to drop, but because the battery power is available, $E_2$ drops only to $E_1$ minus the voltage across $CR_1$, and the circuit continues to function. In this type of circuit, the load must be able to operate adequately on that slightly lower voltage.

### 3.5.7.4 Intermittent Surge Power

The high-rate discharge capability of nickel-cadmium batteries permits their use in applications requiring short-time, intermittent surge power. In such applications the battery is under constant charge. When the periodic need arises for a relatively large amount of power beyond the capability of the power supply, the battery delivers that

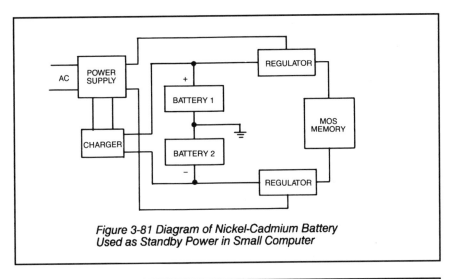

Figure 3-81 Diagram of Nickel-Cadmium Battery
Used as Standby Power in Small Computer

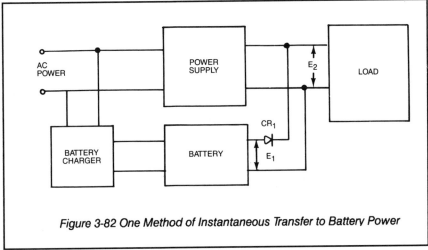

Figure 3-82 One Method of Instantaneous Transfer to Battery Power

energy. This is similar to a capacitor energy storage/discharge function, but differs in that the battery is capable of much longer periods of discharge at higher power levels.

An example of this application is the telephone line carrier equipment that allows a second private line of communications in a location where only one set of telephone lines is available. The communication signal is modulated at the telephone central switching station, transmitted on the one set of lines available, and demodulated at the phone receiver location. The battery is continually charged at low rates over the telephone lines. When the telephone is called, the battery supplies the energy required to ring the bell and sometimes even the energy required for the voice communication.

## 3.5.7.5   Motor Loads

Nickel-cadmium batteries are used in a variety of motor driven devices, for example, electric toothbrushes, shavers, garden tools, power tools, starter motors, tape recorders, and radio-controlled hobby cars. Battery selection requires a thorough understanding of the motor characteristics as well as the gear trains, fans, blades or other power train components. Motor design and battery selection go hand in hand and are discussed in this section.

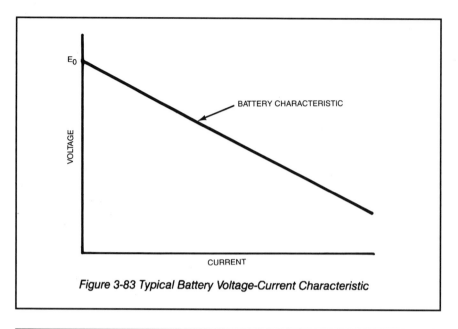

*Figure 3-83 Typical Battery Voltage-Current Characteristic*

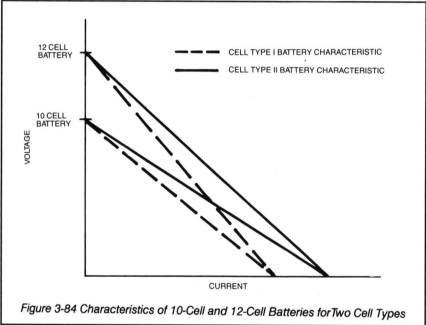

*Figure 3-84 Characteristics of 10-Cell and 12-Cell Batteries forTwo Cell Types*

Trade-offs in battery selection, as in other applications, must include size, weight, run time, performance, and economics. Battery voltage regulation is a major consideration for motor design. In Figure 3-83, $E_o$ is the effective open circuit voltage, and the slope of the line is a result of the effective internal DC resistance ($R_e$) of the battery (See Section 3.1.2).

Battery discharge characteristics depend on the number, type, and size of cells. In general, effective internal resistance is (approximately) inversely proportional to cell capacity (size). However, it can also vary within a cell size depending upon cell design.

When the cells are connected in series the individual resistances will combine to yield the effect shown in Figure 3-84. The battery characteristics also vary de-

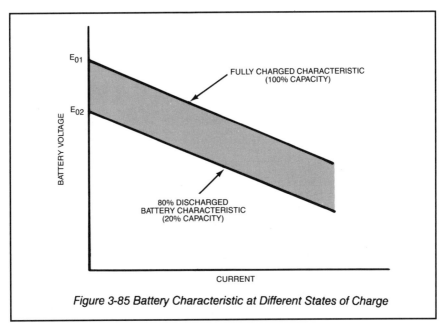

*Figure 3-85 Battery Characteristic at Different States of Charge*

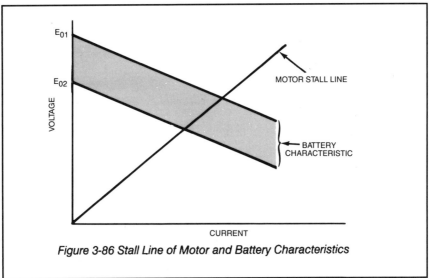

*Figure 3-86 Stall Line of Motor and Battery Characteristics*

pending on the state of charge. Figure 3-85 shows this variation for residual capacities from 100 per cent to 20 per cent.

Figure 3-86 adds the motor stall line characteristic to the battery supply characteristic of Figure 3-85. By adding constant torque curves the full family of characteristics is shown in Figure 3-87. The relationship between torque and speed may now be observed over 80 per cent of the battery capacity and interpreted at any point between the limits.

Trade-offs between operating battery/motor power transfer and the total system run time may need to be considered during the process of matching the battery/motor/load components in a specific system. The characteristics of the motor and any load device may be conveniently combined in the form of input-output diagrams. For example voltage vs. current input characteristics of a motor may be over-

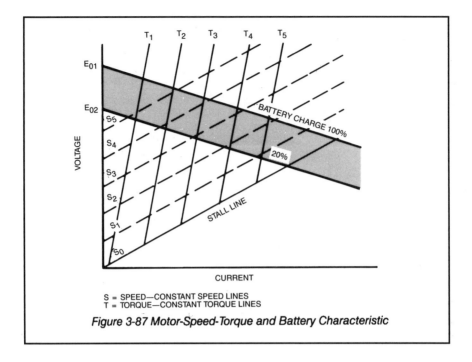

*Figure 3-87 Motor-Speed-Torque and Battery Characteristic*

laid by torque vs. speed output. The characteristics of a load device, such as for example the pressure vs. volume output of a blower or pump, may be substituted directly for torque/speed scales in the case of an additional system element on the output of the motor. The characteristics of a battery under consideration may then be plotted directly on the voltage/current scales to determine system output performance and allow calculation of system run time from the battery capacity at the discharge rate selected. Adjustments in the characteristics of any individual element may then be made and the effects observed in terms of total system performance.

Figure 3-88 illustrates the E vs. I operating points required by a motor in order to extract equal amounts of power from three different battery selections. This curve could be used to match three different motor designs to produce equivalent output torque/speed under the assumption that all three motors are designed to operate at the same efficiency over this range. Calculation of the run time available from each of the three combinations, as result of the three different battery drain rates and probably three different battery capacities at those rates, will then permit selection of the battery/motor combination providing the longest run time under otherwise equivalent system performance.

### 3.5.7.5.1 *Heavy Motor Loads*

Sealed nickel-cadmium batteries are well suited for the heavy motor load applications found in power tools, such as drills. These tools require high levels of power combined with a lightweight battery. Special versions of the nickel-cadmium battery are capable of handling very heavy drain rates for a moderate period of time while maintaining high voltage so that a maximum amount of power is delivered to the load.

Applications with normal operating drain rates of 5C to 20C are common with some power tools requiring as much as 30C for short periods of heavy load that approach motor stall.

The ability of nickel-cadmium batteries to withstand shock and vibration, to be quickly recharged, and to be stored in any state of charge, contribute to making sealed nickel-cadmium batteries the ideal power source for cordless power tools and appliances.

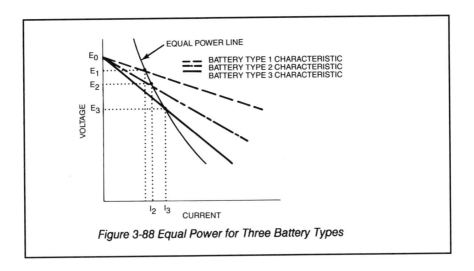

*Figure 3-88 Equal Power for Three Battery Types*

### 3.5.7.5.2 *Light Motor Loads*

Nickel-cadmium batteries are widely used in household products that have operating loads at the 1 to 10C rates such as cordless vacuums, toothbrushes, shavers and mixers. Here the battery provides a small lightweight power source that is kept always ready by maintaining it on charge. The high energy density and ability to withstand continuous overcharge make nickel-cadmium batteries ideal for these cordless appliances.

### 3.5.7.6    **Consumer Devices/Products**

Nickel-cadmium batteries are used to power many consumer devices. Mixers, telephones, flashlights, camera flashes, and portable computers are a few examples.

As most consumer devices are not provided with special low-temperature charge control, the operating instructions should advise the customers not to leave the unit on charge where temperatures may be below freezing, such as in garages and tool sheds during winter.

The consumer should not expect the device to operate before the battery is charged and should be instructed to fully charge the battery prior to use. In some cases this instruction is printed on tape placed over the operating switch where the customer cannot miss it, thus eliminating unnecessary returns of devices perceived as inoperable.

Another potential misunderstanding results when a charger is plugged into an outlet controlled by a switch that is off and the battery does not get charged. This situation can be avoided by incorporating a charging indicator, such as a light-emitting diode (LED), to positively indicate that the battery is receiving charge current.

### 3.5.7.7    **Hobby Uses**

Radio controlled (RC) cars, boats, and planes are in widespread use both as adult hobbies and children's toys.

The transmitter and receiver batteries usually require no special knowledge as they are typically charged and discharged at standard rates. However, in RC hobby cars where the nickel-cadmium battery provides the power for racing, battery knowledge can provide the winning edge. These race car batteries are typically fast-charged and high-rate discharged.

A full charge is essential to finishing the race. As excessive battery temperature reduces charging efficiency, actual cell capacity and voltage, fast-charge control schemes that minimize battery heating while ensuring full charge are best. Therefore

TCO or negative dV/dt systems, possibly with some low-rate topping charge, may provide the best capacity input. Batteries should be allowed to cool after discharge between races before attempting recharge. When not in use, batteries should be stored in a cool environment (15-25°C).

As motor speed is typically directly proportional to battery voltage, a low internal resistance ($R_e$) cell should be chosen to deliver the highest battery voltage at high discharge currents. The $R_e$ of the battery will be the total of all the cells plus the interconnections and lead assemblies. A battery of series-connected cells can only yield the capacity of the lowest capacity cell in the pack.

For maximum performance, the time between charging and use should be minimized to lessen the effects of self-discharge.

The serious hobbyist should be familiar with previous sections on fast-charging, high temperature charge acceptance, $R_e$/voltage, and battery/motor matching. Knowledge, experimentation, and selection of cells with low $R_e$ will yield maximum performance.

### 3.5.7.8    Toys

Nickel-cadmium rechargeable batteries used in toys eliminate the constant incremental cost of new primary batteries, thereby reducing the lifetime cost of the toy.

A second advantage of nickel-cadmium batteries is the nearly constant discharge voltage, resulting in a uniform performance level. This is unlike primary batteries whose voltage decreases as capacity is used, resulting in decreasing performance. The toy using rechargeable nickel-cadmium batteries will not experience a gradual performance decrease. When the device no longer runs, a simple recharge will place it back in service, extending play without battery procurement problems.

Two sets of batteries, one in the toy and one on charge, may ensure that play is not interrupted waiting for batteries to recharge and will essentially allow continued use of the toy.

### 3.5.7.9    Portable Audio/Video

Nickel-cadmium batteries are an ideal choice for portable audio/visual devices as they have the desired features of high energy density, long cycle life, and can be stored in any state of charge, as well as on charge. However, care should be taken that batteries left on overcharge do not experience high temperatures which limit their life.

### 3.5.7.10    Military

Sealed nickel-cadmium cells are used in many military applications including submarines, avionics, missiles, portable infantry communications, detectors, instruments, emergency lights, remote targets, laser designators, as well as common flashlights.

Generally these applications require the ability to operate in environmental extremes, the strength of construction to withstand shock and vibration, and the assurance of excellent quality and reliability.

Many sealed nickel-cadmium cells are manufactured to meet stringent military requirements including first article testing and ongoing piece-part inspections. Typically these cells are shipped with certificates of conformance.

Due to the specialized nature of these applications, performance requirements should be discussed with cell manufacturers at the feasibility study stage.

### 3.5.8    SUMMARY

The preceding discussion illustrates the benefits, cost effectiveness, and usefulness of the nickel-cadmium sealed-cell battery. Highlighted strengths of these batteries are their long operating life, high-rate discharge capability, simple storage, ruggedness, ability to operate in a broad range of environments and positions, and continuous overcharge capability. Careful attention to circuit design and matching the battery to the load will help the designer realize the full potential of nickel-cadmium batteries.

This section has also emphasized the economic advantages of using the nickel-cadmium battery. The long service life of the battery and simplicity of the charger are major considerations in reducing the life-cycle cost of a system.

A diversity of applications, including standby power for alarms, lighting systems, and computers; power for heavy load applications such as in power tools; light-weight power for consumer devices such as shavers; and military uses, have found the sealed nickel-cadmium battery to be the best choice.

# 3.6   Battery Testing, Quality Control, and Specification

The keys to success in designing applications that use batteries are:

- knowing the performance characteristics associated with various batteries,
- understanding how these characteristics will be expressed in the applications's operating environment and
- designing with these distinctions in mind.

Most designers find that application-related testing is an essential part of the design process. Testing, however, can be expensive and should be considered the last-resort after other sources of information (such as this Handbook and consultation with cell manufacturers) have been exhausted. This section provides some suggestions on when battery testing may be advisable and on ways to get the most from the testing.

Battery testing is normally performed for two major purposes:

1) battery qualification — studying how the battery is likely to perform under representative conditions and conducting head-to-head comparisons of different candidate batteries. The results are data that the designer can use in selecting and specifying the battery and designing the supporting equipment.

2) product verification testing — ongoing testing that allows verification of battery quality as part of the manufacturing process.

Battery testing also relates directly to the quality control process and to the generalized specification of sealed nickel-cadmium cells. These topics will also be discussed as part of this section.

## 3.6.1   QUALIFICATION TESTING

Some users of batteries decide to perform special testing to qualify batteries for use in a specific application. This qualification testing usually consists of a series of specific tests in which the battery or cells must meet certain minimum or maximum limits for each test procedure. These qualification tests take on many forms and have various names, such as type and characterization tests. A type test may contain not only initial performance measurement tests but also some form of life exposure with periodic performance measurements.

### 3.6.1.1   Performance Testing

The purpose of performance testing is to define the electrical characteristics of the battery under a variety of conditions. Since sealed nickel-cadmium cells may increase in capacity during the first few cycles for any specific set of conditions, it is necessary to perform stabilization cycles prior to measuring the performance to obtain realistic and repeatable data.

While the details of performance testing are usually determined by the application, some general suggestions on approach are possible. Often the first set of tests to be performed are those that confirm the successful operation of the battery under nominal conditions, i.e. with the battery new, but stabilized; recharged in the optimum manner; and at room temperature. If the battery can not meet the load requirements under favorable conditions, there is no point in continuing. Assuming those tests are favorable, battery operation should be confirmed at the temperature extremes for which the product is expected to operate.

Once the basic ability of the battery to accommodate the load has been determined, then the charge acceptance of the battery can be tested. The ability of the proposed charging system to recharge the battery within the expected application profile is important. As is obvious from the previous sections, charging at the extremes of the temperature envelope is especially important to ensure proper charger function. Often, a charging system that is built to comply exactly with the anticipated application's requirements extracts a penalty in increased cost and complexity and/or a decrease in battery life. This then becomes a situation where careful testing and feedback to those developing the application requirements may pay substantial benefits to the designer. The result can be substantial cost savings through a better understanding of battery behavior.

### 3.6.1.2 Life Testing

Once the performance testing has confirmed battery's ability to meet the basic application requirements, it is time to move on to testing that assesses the battery's life in the application.

Qualification testing is usually conducted on an accelerated basis in one form or another to reduce the test time to a practical length. A life test regime usually attempts to simulate, to varying degrees, the actual expected use pattern of the device. Such regimes may compress calendar time by ignoring the intermittent use patterns and storage (idle) periods which are typical of most applications. Unfortunately, accurately simulating all of the variables encountered in typical real-time use of the product is difficult, if not impossible, in a test. In any attempt to compress test time, extreme care must be taken to avoid any nonlinear life-limiting stresses on the battery which cannot be correlated with the expected real-time use of the battery. Without thoughtful consideration of the significant variables, a life test program may yield misleading conclusions or comparisons.

In most applications, the battery is exercised randomly with some shallow discharges and some deep discharges. The effect of this random usage is typically beneficial for cell performance and may extend the life beyond that found in a fixed, regimented cycle regime such as a qualification test.

With the increase in popularity of fast-charging systems, which can fully charge a battery in one hour or less, cyclic life testing with many cycles per week is quite common and provides a guide to battery cycle life expectancy. However, this test, unless used with fast-charge cells and charged by the same method as the end-use application, may be an additional nonlinear stress not seen in the real application.

Tests designed to generate a data base for defining life characteristics of nickel-cadmium batteries are of two types: real-time life testing or accelerated life testing.

### 3.6.1.2.1 Real-Time Testing

Real-time testing, although the most representative of all testing, has a fundamental limitation. Since the life of sealed nickel-cadmium cells is measured in years, a very long period of time is required to obtain meaningful failure data. The wide range of variables encountered in real-life applications often dictates consideration of a similar broad range of life-testing regimes. Between needing to maintain a statistically significant test sample and needing to test a range of conditions, real-time testing can consume quantities of equipment and manpower for extended periods. The practical result is that real-time testing is rarely used to provide information for design decisions. For such uses accelerated testing is routinely used. However, real-time testing still provides the best reproduction of battery life and can provide an excellent baseline for comparison of accelerated testing data. Many manufacturers maintain ongoing real-time testing programs to accumulate baseline data on cell performance.

### 3.6.1.2.2 *Accelerated Testing*

In performing accelerated testing, it is important that the proper variable be accelerated. For cyclic applications, the critical concern is how many cycles will the equipment last. Here some way must be found to substantially increase the number of cycles per day while maintaining an accurate simulation. In float or standby applications, the number of cycles will be relatively small since the battery spends most of its life on charge. For these applications, some way must be found to accelerate the aging effects of constant overcharge. The best answer now known is temperature stress.

*Cyclic Acceleration*   Accelerated life testing for cyclic applications typically increases the number of cycles per day by eliminating all idle time and most excess overcharge time.

Standard-rate cells designed for 0.1C charging can be life tested to full depth of discharge at cyclic rates no higher than about one cycle per day. This is a rather insignificant cyclic accelerating factor for producing wearout in a reasonable period of time. Quick-charge cells capable of charge rates up to 0.33C can be tested at cyclic rates of about three cycles per day. However, time and temperature may be the limiting factor in either case just as in standby applications. Fast-charge cells designed to tolerate charge rates of 1C or greater can be exposed to accelerated cyclic rates as high as twelve cycles per day. Calendar time is not the life measurement parameter used in the high cyclic rate regimes discussed above, but rather actual cycles to failure. The objective should be to provide the optimum data in the minimum calendar time without introducing unpredictable and unrealistic abusive stress factors.

*Temperature Acceleration*   For accelerated life testing of batteries used in float applications, the test parameter normally used for acceleration is temperature. A great deal of test data and testing experience is necessary to interpret the results of an accelerated float test and use these results for predicting life under normal conditions. The critical assumption in temperature-accelerated testing is that the key cell failure modes are temperature sensitive. The priority wearout failure modes, as discussed in Section 3.4, are shorts and electrolyte loss. The cell components whose design function is to insulate or to seal are all polymeric materials. They inherently deteriorate in time and elevated temperatures are known to speed the process. Thermal-acceleration testing assumes that the degradation rate of these insulators and seals is functionally dependent on temperature through the Arrhenius relationship. With all other factors constant, these degradation rates roughly increase by 20 to 145 per cent for every 10°C rise in cell temperature as illustrated in Figure 3-89. Differences in design and materials account for the rate differences that are reflected in Lines A and B.

The protocol for thermal stress testing calls for overcharging a group of cells at a rate appropriate for the cell design being tested and for the intended operating conditions. The test temperature is selected to be significantly higher than the intended application's. The temperature increase must, however, be limited to the range that does not introduce new failure modes. Accelerated life tests of sealed nickel-cadmium cells are often conducted at temperatures of 37°C, 48°C, 60°C, 72°C and 86°C. The cells are periodically discharged to measure the progressive degradation of cell capability as well as to simulate the anticipated infrequent discharge of such applications. The cell failure criterion typically used is the cell is failed when its capacity falls below 50 per cent of that available at the elevated test temperature when the cell is new.

The results of accelerated life tests are normally represented in terms of characteristic life (that is, the time when 63.2 per cent of the test population at one temperature have failed) as a function of temperature. Such a plot is illustrated in

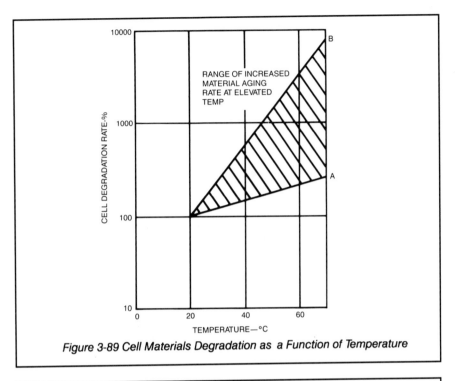

Figure 3-89 Cell Materials Degradation as a Function of Temperature

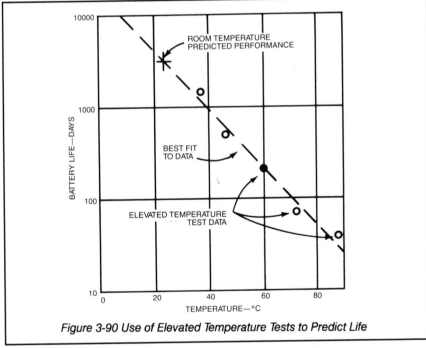

Figure 3-90 Use of Elevated Temperature Tests to Predict Life

Figure 3-90. Normally several groups of cells are put on test at the same time, but at different temperatures. Naturally the higher-temperature cells fail first. The result is a steadily improving estimate of real-condition life as new points become available at the lower temperatures. The temptation, of course, is to dispense with the lower temperatures to speed testing. Without the availability of intermediate condition data, the usefulness of accelerated life tests is, at best, limited to comparison between groups

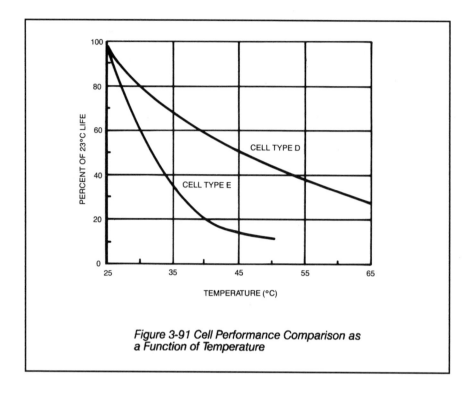

*Figure 3-91 Cell Performance Comparison as a Function of Temperature*

of batteries at the accelerated condition. These comparisons are only valid if the accelerating mechanism has the same effect on both groups of cells. For example a comparison between the cell types in Figure 3-91 could produce seriously misleading results for testing at only one temperature.

## 3.6.2 PRODUCT VERIFICATION TESTS

The qualification testing provides a picture of how the battery should perform. The role of product verification testing is to indicate that the batteries will actually perform as they should. To assist customers in developing an incoming inspection or product verification program, Gates has developed the guidelines presented in Table 3-5. These guidelines are generic; a customized program designed for a specific application normally makes more sense when large volumes of cells or special operating conditions warrant.

## 3.6.3 BATTERY TROUBLESHOOTING

There may be occasions where a cell or battery is suspect. In such cases, Gates has found the standard evaluation procedure shown in flowchart form in Figure 3-92 to be useful. This troubleshooting procedure will normally distinguish between batteries suffering from a temporary failure due to the way they have been utilized and batteries that have permanently failed. The permanent failures can then be evaluated to determine the application parameters that may have resulted in their failure.

## 3.6.4 STANDARDS/REGULATORY AGENCIES

A number of organizations and agencies are involved with the specification and/or assessment of the performance capabilities of nickel-cadmium cells and batteries.

| Character-istic to be Checked | AQL Suggested for Customer | Reason for Test | Equipment Required | Inspection Limit (At Standard Temperature) | Failure Disposition |
|---|---|---|---|---|---|
| OCV | 1–3 Cell Battery    0.65<br>4–9 Cell Battery    1.00<br>10–19 Cell Battery    2.50<br>20+ Cell Battery requires negotiation | Open or shorted cells and broken welds | Voltmeter | For single cell batteries:<br>  OCV = 0.5V<br>  minimum<br>For each additional cell, increase the minimum OCV by 1.2 volts.<br>Example:<br>8 cell battery<br>$0.5 + 7(1.2) = 8.9$ | If OCV is lower than limit, measure individual cell voltages; accept battery if all cells have 0.5 minimum. If cells are not accessible or are less than 0.5 V, apply low voltage test below. Cells may exhibit low OCV if they are accidentally externally shorted or have experienced long shelf life. |
| Capacity | 2.5 | Sufficient run time | Power supply and test fixtures | Per specification requirements at specified temperature | Perform one additional retest to assure proper connections, etc. |
| Battery dimensions | Functional: 1.0 Non-functional: 4.0 | To insure fit | Calipers, micrometers, ht. gage, etc. | Per drawing dimensions | You must insure fit. |
| Weld strength | 2.5 | Weld integrity | Tensile tester | Per specification | Reject |
| Marking, name-plates, date codes | 4.0 | Identification and warranty | Visual inspection | Per specification | Reject |
| Visual workman-ship | 4.0 | Appearance | Visual inspection | Per standard requirements | Reject |

### Standard Low Voltage Test:

1. Charge the battery at the 0.1C rate for 30 minutes.
2. Rest the cell/battery 90 minutes in open circuit condition.
3. Measure the open circuit voltage. If the OCV is equal to or greater than 1.2 volt/cell, the battery is good.

*Table 3-5 Product Verification Reference Guide*

The area of interest of some of these agencies is the product functional performance in specific applications. Other agencies concern themselves with the safety of the product during manufacturing operations, during transportation, during original equipment manufacturer (OEM) assembly, and/or during its use in the end product.

Conformance to the standards of some of these agencies is completely voluntary while others are compulsory, backed by a broad range of governmental authority. Some agencies are concerned only with local issues while others are involved in national and/or international commerce.

These standards can provide a means to communicate from the battery manufacturer to the OEM designer and finally to the product user the performance capabilities of the battery and the cautions to be observed. Some examples of these agencies are ANSI SAE, IEC, UL, CSA, DOT, Factory Mutual, NFPA, and OSHA.

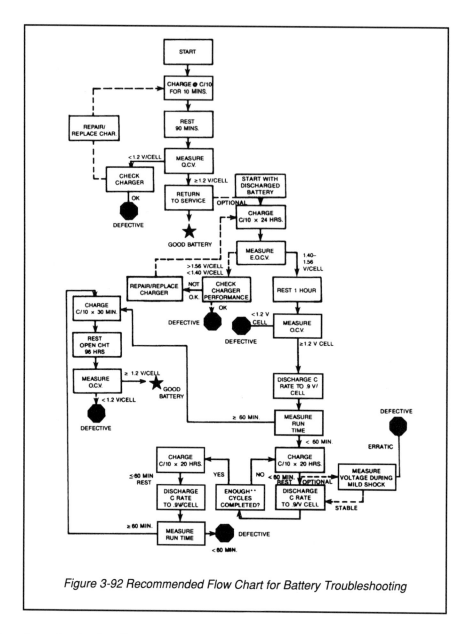

*Figure 3-92 Recommended Flow Chart for Battery Troubleshooting*

### 3.6.4.1 Standards for Sealed Nickel-Cadmium Batteries

The principal American standard which specifies the performance characteristics of sealed nickel-cadmium batteries is the American National Standards Institute, (ANSI), document C18.2 - *Specifications for Sealed Rechargeable Nickel-Cadmium Cylindrical Bare Cells.* This document provides the standardized definitions of designated electrical, mechanical, and dimensional specifications, the numerical values of which are declared by the battery manufacturer. It also contains the test and/or measurement procedures by which the actual performance of the batteries may be confirmed relative to those declared specifications.

Dimensions of nickel-cadmium consumer type batteries, designed to replace ordinary dry cells, are covered in ANSI C18.1 - *Specifications for Dry Cells and Batteries.*

Underwriters Laboratories, Inc. (UL), with the assistance of Industry Advisory Councils, develops Standards for Safety for the safe operation of products which might otherwise present the possibility of a hazard of electrical shock, fire, chemical, casualty, or other nature. Some of these Standards deal directly with the safety of the battery charger, such as:

UL 1236 —Electric Battery Chargers
UL 1310 —Direct Plug-In Transformer Units (Chargers)

Other Standards for safety deal with batteries as only one of several components in the overall safety of battery powered products such as:

UL 45 —Portable Garden Tools
UL 82 —Electric Garden Appliances
UL 478 —Electronic Data Processing Units/Systems
UL 1270 —Radio Receivers, Audio Systems, Accessories

A third area of UL concern is specifying the minimum functional performance of battery powered products, where those products are used to provide safety against the hazards of fire, burglary, explosion, etc. Typical of these Standards, which generally involve battery back-up power, are:

UL 924 —Emergency Lighting and Power Equipment
UL 1023—Household Burglar Alarm System Units

UL provides for continual review and updating of their Standards to keep pace with experience and product developments. They also provide testing and periodic reexamination services to ensure that the products as manufactured will meet the specifications called for in the standards. Products which do meet these minimum criteria are privileged to bear the UL Listing mark or the Component Recognition mark.

The Canadian Standards Association (CSA) provides services in Canada which are similar to those of the UL in the United States. The CSA listing mark, however, is obligatory for commercial sale in Canada, compared to the voluntary use of UL standards and services. The CSA Standards require product characteristics which are, for the most part, completely comparable to UL requirements. Two of the CSA Standards are:

CSA-C22.2 No. 107 Rectifying Equipment (All Chargers)
CSA-C22.2 No. 141 Unit Equipment for Emergency Lighting

Factory Mutual serves a function for industrial products which is similar to the one which UL serves for other products.

The National Fire Protection Association (NFPA) publishes the National Electrical Code (NEC), a standard used extensively to judge construction materials, components and methods. NEC -Article 700 -Emergency Systems specifies the overall performance requirements for emergency lighting systems which is one of the principal bases upon which UL specifies and examines emergency lighting luminaires and the batteries which operate them.

The International Electrochemical Commission (IEC) publishes IEC -285 - 1983 -*Sealed Nickel-Cadmium Cylindrical Rechargeable Single Cells*. This document is similar to ANSI C18.2 in its coverage of characteristics.

Many countries besides the U.S. and Canada have active standard-setting organizations. Among the most active of these organizations, at least in regard to batteries, are VDE in Germany and JIS in Japan. Table 3-6 provides a cross-reference

| Organization → Standard → Generic Cell Designation | ANSI C28.2* | IEC 285 | JIS C8705 | DIN | Dry Cell Reference ANSI | IEC |
|---|---|---|---|---|---|---|
| N | KR115/XXX | KR 12/30 | KR-N | | N | R1 |
| AAA | | KR 11/45 | KR-AAA | | AAA | R03 |
| 1/3AA | KR142/XXX | KR15/18 | KR-1/3AA | | | |
| 1/2AA | KR142/XXX | KR 15/29 | KR-2/3AA | | | |
| AA | KR142/XXX | KR 15/51 | KR-AA | GSZ0.5 | | |
| AA HH | | | | | AA | R6 |
| 5/4 AA | | | | | | |
| 1/3 A | KR160/XXX | | | | | |
| A | KR160/XXX | KR 17/51 | | | | |
| 1/3 $A_f$ | KR168/XXX | KR 18/18 | | | | |
| 2/3 $A_f$ | KR168/XXX | KR 18/29 | | | | |
| 4/5 $A_f$ | | | | | | |
| $A_f$ | KR168/XXX | KR 18/51 | | | | |
| 1/2 $C_s$ | KR222/XXX | KR 23/27 | KR-2/3SC | | | |
| 4/5 $C_s$ | | | | | | |
| $C_s$ | KR222/XXX | KR 23/43 | KR-SC | GSZ1.2 | | |
| $C_s$C | | | | | | |
| $C_s$D | | | | | | |
| 1/2 C | KR257/XXX | | | | | |
| 3/5 C | KR257/XXX | | | | | |
| 2/3 C | KR257/XXX | KR 27/33 | KR-2/3C | | | |
| C | KR257/XXX | KR 27/50 | KR-C | GSZ1.8 | | |
| C HH | | | | | C | R14 |
| 1/2 D | KR334/XXX | KR 35/44 | | | | |
| 2/3 D | | | | | | |
| 3/4 D | | | | | | |
| D | KR334/XXX | KR 35/62 | KR-D | GSZ4 | | |
| D HH | | | | | D | R20 |
| 1 1/4 D | | | | | | |
| F | KR334/XXX | KR 35/92 | KR-F | GSZ7 | | |
| M | | KR 44/91 | KR-M | | | |
| 2 M | | | | | | |

* For ANSI cell standards, the first group of figures in the designation indicates the maximum diameter of the bare cell in units of 0.1 mm. The second group of figures (XXX), the values of which shall be provided by the manufacturer, indicates the maximum overall cell height in the same units.

*Table 3-7 Cross-Reference Listing of Standards for Sealed Nickel-Cadmium Cells*

listing of the designations used by various international and national standards organizations in identifying various sealed nickel-cadmium cell sizes.

## 3.6.5   SUMMARY

Although there are a wealth of resources for technical data on sealed nickel-cadmium batteries, a well-designed qualification test program provides direct assurance that the battery is correctly matched to the application. With help from the manufacturer, such a qualification test program does not necessarily need to be lengthy or expensive. Once the manufacturing program has begun, a routine product verification program or incoming inspection routine will provide assurance of the quality of the incoming product.

# 3.7   Safety

Sealed nickel-cadmium cells and batteries have compiled an excellent safety record. Millions of cells have been produced for use in a diversity of both consumer and industrial products. When properly applied, these cells are a valuable, safe source of electrical energy. However, like any battery, sealed nickel-cadmium cells and batteries present possible hazards if mistreated or misapplied. The following sections discuss the various areas of concern and suggest ways of reducing or eliminating possible problems.

## 3.7.1   GENERAL

**DO NOT PUT IN FIRE OR MUTILATE; MAY BURST OR RELEASE TOXIC MATERIALS. DO NOT SHORT CIRCUIT; MAY CAUSE BURNS.**

Potential hazards may arise from the improper use of sealed nickel-cadmium rechargeable batteries. Manufacturers and assemblers of battery-using products or systems must ensure that the systems are properly designed and that adequate battery handling procedures are in place. Where appropriate, the end-use consumer should be made aware of these potential hazards and ways to avoid them.

## 3.7.2   POTENTIAL BATTERY HAZARDS

In the following sections potential hazards from the improper use of batteries are discussed individually along with methods for reducing or preventing the potential hazard.

### 3.7.2.1   Chemical Burns from the Electrolyte

The electrolyte used in most nickel-cadmium cells is a mixture of potassium hydroxide (KOH) and water to a concentration of approximately 30 per cent. This material is a strong caustic, is classified as corrosive, and can cause significant chemical burns if it touches human tissue. In the event that a sealed cell should leak, protective gloves should be worn when handling the cell. Never rub your eyes if electrolyte gets on hands or fingers. First aid treatment requires that the caustic material be diluted with copious amounts of clean water. Seek medical attention if a significant amount of KOH was involved or if the caustic material touched the eyes.

### 3.7.2.2   Ingestion of Very Small Sealed Cells

Most cells are too large to be swallowed but cells of any size should never be placed in the mouth, nose, or ears. Damage to tissue may result from chemical and/or electrical burning. A physician's care should be sought in all cases of ingestion or where the cell becomes lodged in the nose or ears. In all cases of ingestion, the progress of the cell through the body should be carefully monitored and surgical intervention is usually indicated if the progress is stopped. Cells lodged in the nose or ears should be immediately removed by the physician.

### 3.7.2.3   Burns or Excessive Heat from High-Rate Discharge

The sealed nickel-cadmium cell can deliver power at high rates. Typical maximum discharge rates for these cells if short-circuited can be as high as 100C. Accordingly, the rate of power delivery should be adequately controlled by the end-product using this power source.

Furthermore, care must be exercised in the handling and use of the cell to avoid external shorts. Accidental short circuiting may be caused by a number of conditions: a) placing an uninsulated multi-cell battery on a metal shelf or bench; b) using uninsulated tools when working in the vicinity of the battery; c) wearing rings, metal watch bands, I.D. bracelets, and other jewelry without suitable insulating protection; or d) carrying coins or keys in your pocket with a battery.

High-voltage batteries consisting of many cells in series have the potential to arc along paths which have become contaminated. This arcing can result in a carbonized path, eventually resulting in the equivalent of a short.

The very high-rate short-circuit current that may flow into an external short or along a carbonized path can cause significant heating along the current path. Terminals, cell interconnections, or lead wires may become very hot with the potential for personnel injury or ignition of adjacent flammable materials. The design of end-products must consider this power delivery capability. For example, a current-limiting device such as a fuse, resistor, diode, or circuit breaker, may be used in the discharge circuit to prevent short-circuit currents. Use of polarized connectors designed to eliminate simultaneous contact of both battery terminals is one excellent way of eliminating most short-circuit hazard.

### 3.7.2.4   Electric Shock

Batteries consisting of more than 30 cells in series present voltages greater than 43.5 volts when on charge. This voltage is generally considered the threshold of electrical shock capability for direct currents. Appropriate design measures must therefore be taken, warning labels provided as appropriate, and the batteries handled carefully and in conformance with applicable regulations.

### 3.7.2.5   Proper Disposal

Do not mutilate batteries, as corrosive electrolyte can be released (see 3.7.2.1). Do not dispose of sealed nickel-cadmium cells in a fire, as they may burst explosively or release toxic fumes. Disposal of cells in a charged condition is not recommended for the reasons set forth in 3.7.2.3. In disposing of batteries give appropriate consideration to the total amount of toxic materials in the lot and conform to applicable local, state, and federal regulations.

### 3.7.2.6   Venting

Nickel-cadmium cells can produce hydrogen and oxygen which under some conditions may be vented from the cell. A mixture of hydrogen and oxygen or hydrogen and air will explode if ignited. Mixtures from approximately 5 per cent hydrogen to over 90 per cent hydrogen are explosive. Maximum explosion pressure multipliers of approximately 9 times may be realized with the most explosive mixture (stoichiometric) of 66 per cent hydrogen and 33 per cent oxygen. Pressure multipliers decrease as the hydrogen content is diluted or enriched from that value. Approximately 0.3cc of $H_2O$ will electrolyze to gas per ampere-hour of electrolysis input and yield about 415cc of $H_2$ and 207cc of $O_2$ at one atmosphere. Cells have roughly 2cc of water per ampere-hour of rated capacity. Although sealed nickel-cadmium cells are far safer than flooded batteries, potential ignition sources still should not be allowed in the vicinity of batteries in overcharge.

Most sealed nickel-cadmium cells will not vent sufficient amounts of gases to develop an explosive environment unless the gas is trapped. Precautions must be taken to provide adequate ventilation for any compartment (such as a plastic case) enclosing the battery. **Completely sealed battery compartments are not acceptable; ventilation of some sort must be provided.** This is particularly true for fast-charge cell applications where failure to terminate the fast-charge current would cause rapid venting of the cell.

In the highly unlikely event that the safety vent of a sealed cell becomes inoperative while subjected to abuse, the internal pressure could become sufficient to separate the cover from the can under some conditions. The remote possibility of such an occurrence can be further reduced by minimizing the occurrence of abusive situations which could damage the vent mechanism or generate excessive gases.

### 3.7.3 INTEGRATION WITH THE PRODUCT

Batteries are used in many products designed for use in emergency situations where reliability of the product is very important. The product designer should use great care in integrating the battery into the product and in selecting the appropriate size and type of battery. The designer must also consider the inherent characteristics and performance of the battery.

Thorough instructions on proper use, storage, and charging practices should be provided to the user. A periodic test procedure is advisable as well as a warning that will unmistakably indicate when the battery is no longer functional, i.e. when its condition or performance is no longer adequate for the application.

### 3.7.4 DETACHABLE CHARGERS

When a detachable charger is disconnected from a battery, or device, the terminals of the battery may be exposed and could be shorted by pencil tips, paper clips, coins, etc., resulting in a rapid high-current discharge. This problem may be eliminated by use of the precautions of 3.7.2.3.

### 3.7.5 SUMMARY

Improper use, handling, or disposal of nickel-cadmium cells can pose hazards. Batteries should not be put in a fire nor mutilated or abused, especially in ways that would release their contents. In handling batteries or applying them to devices, care should be taken to avoid any short circuit and to otherwise limit any high-rate discharge.

#### WARNINGS

The WARNINGS set forth below should generally be communicated to each ultimate user of nickel-cadmium batteries (cells). They should appear in an appropriate and effective location for each end product.

- **DO NOT** incinerate or mutilate; may burst or release toxic materials.
- **DO NOT** short circuit; may cause burns.

*Section* $4$

# Sealed Lead Cells
# and Batteries

Starved-electrolyte sealed-lead batteries are the most advanced form of lead battery in use today. Because of their demonstrated superior performance and reliability, these batteries are essential components of a variety of products ranging from inexpensive toys and consumer products to telecommunications systems and aircraft. The features that make starved-electrolyte sealed-lead cells and batteries the choice for such diverse applications are briefly described below.

## FEATURES AND BENEFITS

Among the areas where this form of lead battery offers advantages are the following:

### Excellent Performance

Performance remains the key concern of many designers. By taking advantage of the superior performance of the starved-electrolyte sealed-lead design, they can often use smaller batteries with resulting savings in weight and volume throughout the system.

**Discharge Currents** — The thin-plate, low-impedance design of many sealed-lead cells and batteries gives much higher discharge currents than traditional designs of the same rating. Currents as high as 12C are available at usable voltages.

**Power Density** — Thin plates and minimum amounts of electrolyte mean more of the battery's weight and volume may be devoted to active materials.

**Low-Temperature Performance** — No battery likes low temperatures for discharge, but the starved-electrolyte design minimizes cold weather problems.

**Voltage Maintenance** — The decline in voltage from the beginning to the end of discharge has often caused problems for application designers using classic lead batteries. With starved-electrolyte sealed-lead cells and batteries, the voltage delivery characteristic is very good, especially at higher discharge rates and at temperature extremes. Many of the design compromises necessary due to voltage variation may be eliminated with these batteries.

### Long Life

Starved-electrolyte sealed-lead batteries have demonstrated an enviable longevity whether they are used in float or cyclic duty or just kept in storage.

**Float** — These batteries, when properly charged, offer up to 10 years' life at room temperature before their capacity drops to 80 per cent of its rated value.

**Cyclic** — Starved-electrolyte sealed-lead batteries outperform other lead batteries when used in cyclic service.

**Storage** — The self-discharge rate for many starved-electrolyte sealed-lead batteries is very low. This not only makes storage easier, but products are much more likely to come from storage with some residual capability available. This is especially important in some consumer applications where the purchaser expects to see the product operate straight out of the box, even before he has charged it.

## Simplified Charging

Trying to find a charging scheme that would bring the battery to full charge in all conditions without damaging overcharge used to be a major concern. The result was often unsatisfactory. The gas recombination within the starved-electrolyte battery greatly improves its ability to accommodate overcharge, giving designers a new freedom in tailoring the charger to the application.

**Float Charging** — Charging batteries on float is a sensitive problem—too little charging and the battery will not discharge properly when next needed, but too much charging can shorten the battery's life. Starved-electrolyte sealed-lead batteries may be charged either by constant-voltage or two-step constant-current methods to obtain maximum life while providing adequate discharge performance.

**Fast Charging** — Because of their construction, starved-electrolyte sealed-lead batteries will accept high charge currents without the water loss that curtails the life of other lead batteries. By using a constant-voltage charger set at the proper voltage, fully discharged batteries can be brought back to a high level of charge (80-90 per cent) in less than an hour.

## Design Flexibility

Conventional lead-acid batteries are normally unwelcome guests—tolerated because of their usefulness, but disliked because of their nasty habits. Many applications require that conventional batteries be isolated in separate compartments made from acid-resistant materials with independent venting and drainage systems. Starved-electrolyte cells and batteries do not require this special treatment. Since spillage and corrosion are not problems, these batteries do not have to be separated from other equipment. In fact even sensitive applications, such as computers and aircraft, now locate starved-electrolyte batteries among other electronic equipment. And, these batteries no longer have to be mounted vertically; allowing equipment designers enhanced design flexibility.

## Elimination of Maintenance

Inexpensive batteries that require routine maintenance often turn out to be no bargain at all. It is easy to spend far more than the initial cost of the battery in labor cost for regular upkeep over the life of the battery. Plus, equipment that requires maintenance may be incorrectly maintained. If the battery has to be accessible for regular mainte-nance, the designer loses some latitude in choosing a location for it. Freedom from all of these liabilities is among the reasons that the maintenance-free starved-electrolyte products have made such a strong showing in remote and distributed applications.

## Ruggedness

Starved-electrolyte sealed-lead batteries are tough. Not every one is concerned that the battery continue to perform for a short period of time (and not leak) when it is punctured by shrapnel. But the military services do care. This is one reason that they have selected starved-electrolyte batteries for many of their applications. On a more mundane level, consider electric-start walk-behind lawn mowers. This seemingly innocuous application is a torture test for batteries. Batteries are often located in hot, high-vibration environments. Their use can be sporadic, but intense. The charging systems are relatively crude. And, the end-user knows and cares little about proper treatment of the battery. It is a great testimonial to starved-electrolyte sealed-lead batteries' ruggedness that they are the battery of choice for this demanding application.

# APPLICATION EXAMPLES

With the attributes described above, it is little surprise that starved-electrolyte batteries have found uses in a wide array of equipment. Some examples include:

## Standby Power
- Telecommunications Equipment
- Security Alarm Systems
- Emergency Lighting
- Computers
- Medical Equipment
- Uninterruptible Power Supplies

## Engine Start
- Lawn Mowers

## Portable Power
- Portable Lighting
- Cordless Appliances
- Toys
- Cellular Phones

## Alternate Power
- Computers
- Consumer Electronics
- Instrumentation

# SECTION CONTENTS

The remainder of this section provides information useful in properly selecting the correct starved-electrolyte battery for an application and then designing that battery into the system. For most effective use of sealed-lead cells and batteries, the reader is strongly encouraged to read Section 4 in its entirety. It begins with a discussion of discharge performance in Section 4.1 since this information is vital to selecting the proper battery for the application. Then once the battery is selected, it must be charged correctly so that it will supply the needed discharge performance without adversely affecting its life. Suggestions for tailoring charging to the application are provided in Section 4.2. Since nearly every battery is stored at some point in its life, Section 4.3 explains how to store sealed-lead batteries and cells. Understanding the tradeoffs that affect battery life is the theme of Section 4.4. Section 4.5 provides a variety of applications information on sealed-lead products. Included is a discussion comparing the economic benefits of batteries, information on packaging and location options, operating environment considerations in using sealed-lead batteries, and a brief discussion of typical applications. Since testing is often the only way to understand how a battery will perform in a specific application, Section 4.6 describes some approaches to battery testing and specification. Finally, Section 4.7 covers the safety precautions that apply to sealed-lead products.

# NOTE TO THE READER

Small lead-acid batteries for industrial and consumer applications are supplied in a variety of forms. In general, the sealed, starved-electrolyte versions of these batteries are the most advanced and probably the most common. But, because of the diversity of design techniques used to produce these batteries, there are greater variations among manufacturers or even among product lines than typical of sealed nickel-cadmium cells. The information presented in Section 4 and Appendix B has been developed by Gates Energy Products with specific reference to its line of starved-electrolyte sealed-lead batteries. This information is believed to be generally representative of the performance of other manufacturers' starved-electrolyte sealed-lead batteries. However, all product designers should verify performance information with the battery manufacturer prior to committing to a design.

The material in Section 4 is intended to describe in general terms the performance of starved-electrolyte sealed-lead cells and batteries. In most cases, little quantitative information is provided by the text or figures presented in Section 4. Instead, Appendix B is designed to complement Section 4 by providing up-to-date, quantitative performance data pertinent to current starved-electrolyte sealed-lead battery production from Gates Energy Products. As appropriate, the information in Appendix B may be provided for all production versions or may be specific to one cell or battery size. Both Section 4 and Appendix B focus on the electrical performance of sealed-lead cells and batteries. Physical design data (dimensions, electrical terminations, etc.) for sealed-lead cells and batteries are readily available from the manufacturers.

# 4.1   Discharge Characteristics

The purpose of including a battery in a product design is to obtain electrical current from it. In using a battery, a product designer normally has two questions or concerns: 1) How long will the battery supply the current needed by the product? and 2) How will the voltage behave over the course of the discharge? This section discusses how the current and voltage supplied by sealed-lead batteries vary in response to a wide range of load-related and environmental conditions.

A significant design advantage of starved-electrolyte cells and batteries is their versatility in discharge performance. One product design provides superior performance in applications ranging from starting engines to providing memory backup for computer equipment. Thus battery users may use the same battery to handle widely varying product load scenarios.

## 4.1.1   GENERAL

Before getting into the specifics of discharge performance, some general comments on how batteries perform on discharge are pertinent.

### 4.1.1.1   Discharge Types

In talking about battery capacity and discharge performance, it is sometimes useful to compare a battery to a jar containing molasses. Extracting power from a battery is like turning the jar upside down—you can get a lot out very quickly. But in both cases, there is a significant residue that does not come out in the first rush. To totally deplete either they must be allowed to trickle discharge for many hours. With a battery, it is important to remember that the performance differences between a quick discharge and a long, slow, total discharge may be quite significant.

There are three general classes of discharges for which sealed-lead batteries are typically applied. Each one of them has its own design considerations and each serves substantially different forms of applications. The differentiating parameter is the rate of discharge—whether it is high, medium, or low. Some considerations regarding each category will be presented below.

#### 4.1.1.1.1 *High-Rate Discharges*

Typically high-rate discharges are described as anything above 4C. The primary application of interest here is starting engines where the discharge rate requirement may be quite high (over 10C). The discharges normally last only a few seconds each, although there may be several pulses in a train. Certain appliance applications may also have discharge rates that approach the lower end of the high-rate category.

#### 4.1.1.1.2 *Medium-Rate Discharges*

Stepping down from the high-rate applications, there is a family of applications clustered around the 1C rate. Among the products that often need a battery that is good for a half hour to about two hours are many portable appliances, backup power for alarm and emergency lighting, and uninterruptible power supplies. In many respects, these are the easiest discharges for the battery to handle, neither too high nor too low.

### 4.1.1.1.3 Low-Rate Discharges

Low-rate applications are those with a discharge rate below 0.2C, i.e. applications that require the battery to last more than about five hours. This may be anything from an instrument that is required to operate for an eight-hour shift to microprocessor memory holdup that must provide current for a week or more. These discharges may remove essentially all the capacity and thereby place great strain on a battery.

### 4.1.1.2 Design for Discharge Performance

Any battery is the result of a multitude of design compromises, many of which may affect its behavior during discharge. To understand how discharge performance varies with changing loads or changes in the surrounding environment, it helps to understand a little about discharge mechanisms. In particular, the discharge process has two components: an early phase and a long-term phase.

The early phase is dominated by surface reactions. There is not time for transport mechanisms to have much impact, so all the activity is concentrated at the interface between the plate active material and the electrolyte. To maximize short-term response, i.e. that needed for high-rate performance, the plate surface area per unit volume should be a maximum. Advanced sealed-lead batteries use thin plates increasing the surface area available for reaction within a given volume. The result is enhanced high-rate performance.

For long-term response, i.e. the response to a deep, slow discharge, surface effects are less important. Here there is time for transport mechanisms to come fully into play. This means all of the active materials, not just the surface layer, may be involved in the reactions. The key parameter in determining deep-discharge performance is the weight of active material per unit volume of battery. Starved-electrolyte sealed-lead batteries obtain superior performance in deep discharge through elimination of excess electrolyte which increases the proportion of the battery's weight devoted to other active materials. The result is energy densities which give good performance in deep cycle applications.

## 4.1.2 MEASURES OF DISCHARGE PERFORMANCE

The discharge parameters of concern are cell (or battery) voltage and capacity (the integral of current multiplied by time). The values of these two discharge parameters are functions of a number of application-related factors as described in this section. The general shape of the discharge curve, voltage as a function of capacity (or time if the current is uniform), is shown in Figure 4-1. The discharge voltage of the starved-electrolyte sealed-lead battery typically remains relatively constant until most of its capacity is discharged. It then drops off rather sharply. The area of relatively constant voltage is called the *voltage plateau*. The flatness and the length of this plateau relative to the length of the discharge are major features of these sealed-lead cells and batteries. The point at which the voltage leaves the plateau and begins to decline rapidly is often identified as the *knee of the curve*.

The discharge curve, when scaled by considering the effects of all the application variables, provides a complete description of the output of a battery. Differences in design, internal construction, and conditions of actual use of the battery affect one or both of these performance characteristics (voltage or capacity).

The remainder of Section 4.1 will define the discharge curve in terms which will allow the construction of a complete discharge curve (voltage vs. both capacity and run time) for any cell (battery) proposed for an application using variables and parameter values appropriate to that application.

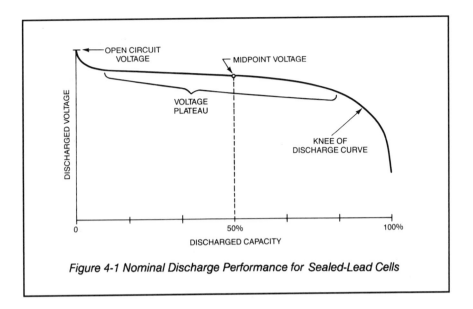

*Figure 4-1 Nominal Discharge Performance for Sealed-Lead Cells*

### 4.1.2.1   Capacity Stabilization

One consideration in evaluating the discharge performance of some sealed-lead cells and batteries is the early rise in capacity. Part of the manufacturing process for all lead cells or batteries is the conversion of the pastes on the electrodes into the active materials needed for successful operation of the cell. This final step in the manufacturing process, called *formation*, does not normally proceed to completion; some of the paste remains unconverted. The early use of the battery completes the conversion process as the battery is charged in service. The result is growth in battery capacity until the battery *stabilizes* at a level that may actually be greater than 100 per cent of the nominal capacity. Unless noted, all results presented in Section 4 and Appendix B refer to stabilized values.

### 4.1.3   BATTERY CAPACITY

The first question that any designer is likely to ask about a battery is "Will it power my product?" This question is usually then refined somewhat: "Will the battery provide adequate current (or adequate power) for the intended length of operation for the product?" Only after these questions about the capacity of the battery have been answered affirmatively, are other concerns (about voltage maintenance, etc.) voiced. Thus, this section first presents information on battery capacity under varying conditions and then moves into more detailed discussions of equivalent circuits, voltage behavior and so on.

   The capacity delivered by a cell is the integral of current (electron flow) over time which equates to the gross number of electrons supplied by the cell to the outside circuit. The number of electrons that the cell will supply is a function of both how the cell is used (current flow, charge method, duty cycle) and the environment in which it is used, i.e. operating temperature. This section will refine the various definitions of capacity and then describe the parameters that affect the capacity a cell will deliver. Capacity is generally measured in terms of ampere-hours or some other current-time product. Knowing the behavior of the cell's voltage under discharge, this capacity translates easily to watt-hours, volt-ampere-minutes or some other measurement of the amount of energy that the cell can deliver to the load.

| | |
|---|---|
| Standard Conditions | = Laboratory Conditions: charge/rest/discharge rates/voltage/temperature |
| Standard Capacity | = Cell capacity measured under standard conditions. |
| Rated Capacity | = The minimum standard capacity. |
| Actual Capacity | = Capacity of a fully charged cell measured under non-standard conditions except standard end of discharge voltage (EODV). |
| Retained Capacity | = Capacity remaining after a rest period. |
| Available Capacity | = Capacity delivered to a non-standard EODV. |
| Dischargeable Capacity | = Capacity which a cell can deliver before it becomes fully discharged. |

*Table 4-1 Capacity Terminology Definitions*

The energy that may be obtained from a sealed-lead cell is dependent primarily upon the discharge current rate, the temperature of the cell, and the conditions under which the cell was charged. Essentially, the charging conditions determine the amount of energy stored within the cell, while the discharging conditions determine how much of that energy is accessible for discharge. The effects of these parameters will be discussed later in this section.

### 4.1.3.1   Battery Capacity Definitions and Ratings

Battery or cell capacity simply means an integral of current over a defined period of time.

$$\text{Capacity} = \int_{\Delta t} i \, dt$$

This equation applies to either charge or discharge, i.e. capacity added or capacity removed from a battery or cell. Although the basic definition is simple, many different forms of capacity are used in the battery industry. The distinctions between them reflect differences in the conditions under which the capacity is measured. Commonly used capacity terms are introduced in Table 4-1 and summarized below.

*Standard capacity* measures the total capacity that a relatively new, but stabilized production cell or battery can store and discharge under a defined standard set of application conditions. It assumes that the cell or battery is fully formed, that it is charged at standard temperature at the specification rate, and that it is discharged at the same standard temperature at a specified standard discharge rate to a standard end-of-discharge voltage (EODV). The standard end-of-discharge voltage is itself subject to variation depending on discharge rate as discussed in Section 4.1.7.

Since cells (or batteries) coming from production may have slight variations in capacity, the value of standard capacity may lie anywhere within the statistical distribution of capacity as manufactured. Figure 4-2 illustrates a typical capacity distribution. *Unless otherwise stated, this Handbook uses standard capacities.*

When any of the application conditions differ from standard, the capacity of the cell or battery may change. A new term, *actual capacity,* is used for all nonstandard conditions that alter the amount of capacity which the fully charged new cell or

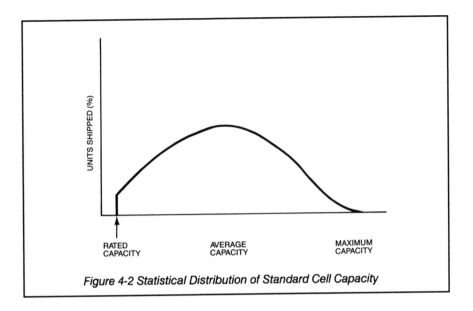

*Figure 4-2 Statistical Distribution of Standard Cell Capacity*

battery is capable of delivering when fully discharged to a standard EODV. Examples of such situations might include subjecting the cell or battery to a cold discharge or a high-rate discharge.

That portion of actual capacity which can be delivered by the fully charged new cell or battery to some nonstandard end-of-discharge voltage is called *available capacity*. Thus, if the standard EODV is 1.6 volts per cell, the available capacity to an end-of-discharge voltage of 1.8 volts per cell would be less than the actual capacity.

Cells and batteries are rated at standard specified values of discharge rate and other application conditions. *Rated capacity* (**C**) for each cell or battery is defined as the minimum standard capacity to be expected from any example of that type when new but fully formed and stabilized. The rated value must also be accompanied by the hour-rate of discharge upon which the rating is based (e.g. 1 hr, 5 hr, 10 hr, 20 hr, etc). The rated capacity for each sealed-lead cell and battery type produced by Gates is indicated in Appendix B.

Rated capacity is always a single specific designated value for each cell or battery model (type, size and design), as contrasted with the statistically distributed values for all other defined capacities. Thus a group of D cells with a rated capacity of 2.5 amp-hours might have standard capacities ranging from 2.5 to 3.0 amp-hrs with an average of 2.65. The Gates process for the manufacture of sealed-lead cells (and batteries) produces a comparatively tight spread in the overall distribution of standard capacity as shown in Figure 4-2.

Figure 4-2 refers to single-cell capacity and NOT multi-cell battery capacity. In any multi-cell battery, the lowest capacity cell in the battery determines its capacity. The distribution of battery capacity, therefore, has the same minimum value as in Figure 4-2 (rated capacity), but its maximum capacity may be somewhat reduced. This reduction depends on the number of cells in the battery and the width (statistical variance) of the capacity distribution of the particular population of cells from which the batteries are actually constructed. If only identical capacity cells were used within each battery, the distribution of battery capacity would be the same as the distribution of cell capacity.

If a battery is stored for a period of time following a full charge, some of its charge will dissipate. The capacity which remains that can be discharged is called *retained capacity*. Section 4.3 discusses storage and its effect on capacity.

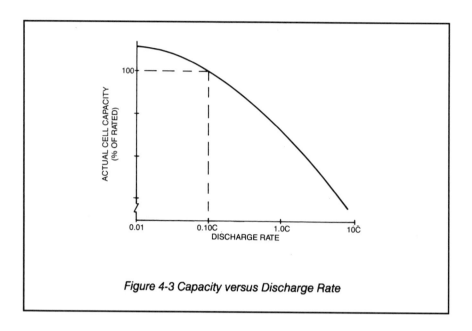

*Figure 4-3 Capacity versus Discharge Rate*

### 4.1.3.2  **Measurement of Fully Charged Capacity**

The capacity of a cell, or battery, is normally measured by completely discharging it while integrating the current over the period of the discharge. Variations in the capacity measurement procedure can result in data inconsistencies.

The most common method of measuring capacity is to discharge the battery with a constant-current load. The load circuit adjusts to maintain a constant discharge current as the battery voltage declines. Recording battery voltage versus time results in a discharge curve similar to Figure 4-1. Calculation of discharged battery capacity is thus only a multiplication of the time needed to reach the specified end-of-discharge voltage (EODV) times the current. An added refinement is the simultaneous use of a current integrator or current shunt to ensure that the load is stable and accurate in maintaining the constant current. A variety of packaged loads designed specifically for constant-current discharges are available for capacity measurement.

An older, less common, and less accurate method of measuring capacity is to place a fixed resistance load across the battery terminals and monitor the voltage as a function of time as the battery discharges. With a fixed resistance, the current decreases as the battery voltage declines. A recorder is used to record the voltage drop across the resistor. The discharge recording of resistor voltage drop is translated to current and then manually integrated over time to calculate the discharged capacity. Use of a current integrator in the circuit can speed the capacity measurement. Unfortunately, the discharge current, which influences actual battery capacity, is variable in this procedure. Thus, relating the results to other application conditions can be quite difficult.

Measured battery capacity depends also on the end-of-discharge voltage used in the measurement. For most accurate results in measuring total battery capacity, the voltage used to terminate the discharge should be below the knee of the discharge curve. This simply means that the end of the discharge should occur after the battery has left the flat plateau of the discharge curve and the voltage is falling rapidly.

Higher values of EODV, when used in measurement procedures, may decrease the accuracy of the results. For EODV's on the voltage plateau, for example, voltage

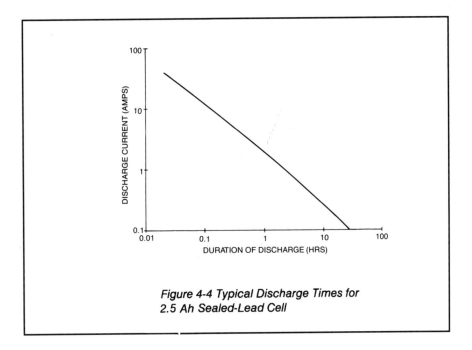

*Figure 4-4 Typical Discharge Times for
2.5 Ah Sealed-Lead Cell*

is dropping slowly with time, so small errors in measured voltage may result in significant errors in the time (capacity) to the end of the discharge. Once the battery is off the plateau, the voltage falls very rapidly and the remaining effective capacity is slight so there is little capacity difference between different EODV's.

### 4.1.3.3  Capacity as a Function of Discharge Rate

The rate at which current is drawn from a battery affects the amount of energy which can be obtained. At low discharge rates the actual capacity of a battery is greater than at high discharge rates. This relationship is shown in Figure 4-3. See Section 4.1.3.7 for a more detailed discussion of capacity ratings.

The information from Figure 4-3 can be used to create a valuable curve of run time versus discharge rate such as shown in Figure 4-4. This shows the amount of time that a certain size of sealed-lead cell or battery will support a given discharge current at room temperature. The data presented in this chart should be regarded as nominal performance for a fully charged battery that has been stabilized at full capacity. Differing conditions, either relating to the battery or the environment, can affect these nominal values.

### 4.1.3.4  Capacity as a Function of Battery Temperature

Starved-electrolyte sealed-lead batteries may be discharged over a wide range of temperatures. They maintain adequate performance in cold environments and may produce actual capacities higher than their standard capacity when used in hot environments. Note that the discharge temperature of concern is that experienced by the active materials within the battery. The time required for a battery to come to thermal equilibrium with its environment may be significant.

Figure 4-5 indicates the relationship between capacity and cell temperature. Actual capacity is expressed as a percentage of rated capacity as measured at 23°C. Quantitative derating curves for the effects of non-standard discharge temperature are presented in Appendix B.

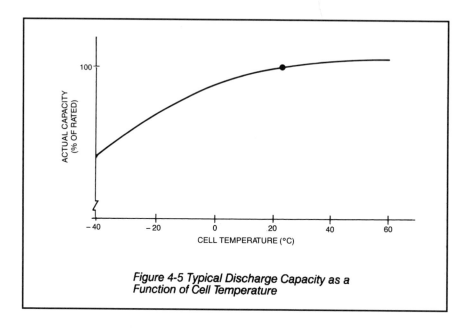

*Figure 4-5 Typical Discharge Capacity as a Function of Cell Temperature*

### 4.1.3.5   **Capacity During Battery Life**

The initial actual capacity of sealed-lead batteries is almost always lower than the battery's rated or standard capacity. However, during the battery's early life, the actual capacity increases until it reaches a *stabilized* value which is usually above the rated capacity. The number of charge-discharge cycles or length of time on float charge required to develop a battery's capacity depends on the specific regime employed. Alternatively if the battery is on charge at 0.1C, it is usually stabilized after receiving 300 per cent (of rated capacity) overcharge. The process may be accelerated by charging and discharging at low rates.

Under normal operating conditions the battery's capacity will remain at or near its stabilized value for most of its useful life. Batteries will then begin to suffer some capacity degradation due to their age and the duty to which they have been subjected. This permanent loss usually increases slowly with age until the capacity drops below 80 per cent of its rated capacity, which is often defined as the end of useful battery life. Figure 4-6 shows a typical representation of the capacity variation with cycle life that can be expected from sealed-lead batteries.

Section 4.4 discusses in more depth the amount of time or number of cycles that can be expected prior to end of useful life.

### 4.1.3.6   **Effect of Pulse Discharge on Capacity**

In some applications, the battery is not called upon to deliver a current continuously. Rather, energy is drawn from the battery in pulses. By allowing the battery to "rest" between these pulses, the total capacity available from the battery is increased. Figure 4-7 presents typical curves representing the voltage delivered as a function of discharged capacity for pulsed and constant discharges at the same rate. For the pulsed curve, the upper row of dots represents the open-circuit voltage and the lower sawtooth represents the voltages during the periods when the load is connected. The use of discharged capacity as the abscissa eliminates the rest periods and shows only the periods of useful discharge. Because each application is unique, individual test-ing should be performed to evaluate the relative capacity gain of pulse discharge compared with continuous discharge.

The significant difference between total discharge capacity values for pulsed

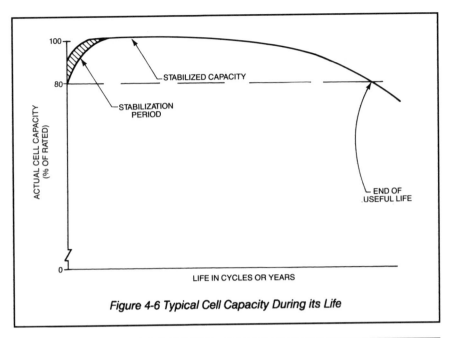

*Figure 4-6 Typical Cell Capacity During its Life*

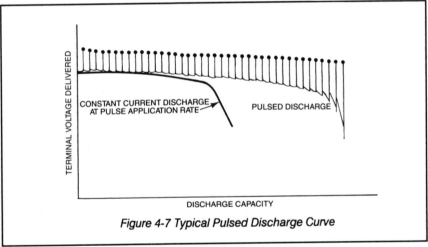

*Figure 4-7 Typical Pulsed Discharge Curve*

and for steady current discharge is caused by a phenomenon known as *concentration polarization*. When current is delivered by the battery, the active material in the plate interacts with the electrolyte to reduce the concentration of the acid in the immediate vicinity of the plate. Since the amount of electrolyte available in the plate pores is less than that required for complete discharge, the delivered capacity at continuous high rate will be limited. However, when time is allowed for the acid to diffuse from the separator back into the pores of the plate, such as during the rest period when pulse discharging, the overall capability to deliver energy is increased.

### 4.1.3.7   Battery Capacity Ratings vs. Discharge Rate

Battery performance may be rated differently depending on battery type and application. This can be confusing for the designer trying to find a suitable battery for a specific application. Most of the confusion centers around the discharge rates used to specify the capacity of a cell or battery and relates to the fact that the deliverable capacity varies inversely with the discharge rate.

| Time for Full Discharge (Hours) | Actual Capacity (10-Hour Rate = 100%) |
|---|---|
| 0.2 | 44% |
| 1.0 | 72% |
| 5.0 | 92% |
| 10.0 | 100% |
| 20.0 | 108% |

*Table 4-2 Typical Capacity Variation At Different Discharge Rates*

Some batteries are rated at the one-hour rate, some at the five-hour rate, some at the 10-hour rate while many specify capacity at the 20-hour discharge rate.

Table 4-2 shows the nominal capacity of sealed-lead batteries at a variety of different rates. Notice that the batteries at the 20-hour rate have about 8 per cent more capacity than at the 10-hour rate.

If all batteries were rated on the same basis, it would be easier to compare one type against another by just considering the data on the battery label. But even then, the relative performance differences between two battery types at one set of conditions may well be different from their relative performance at another set of conditions. Thus, even though battery ratings are a convenient shorthand, the only reliable way to select a battery is to examine actual performance data for candidate batteries at the desired application conditions.

## 4.1.4    CELL EQUIVALENT DISCHARGE CIRCUIT

A battery, unlike many electrical energy sources, has a variable source voltage as well as internal losses that impact the voltage available to the external circuit. The Thévenin equivalent-circuit model is a helpful aid in understanding the discharge capabilities of a cell and how these capabilities may vary. Figure 4-8 shows the equivalent-circuit diagram for a sealed-lead cell.

When a load is connected to the cell terminals, current will flow from the cell

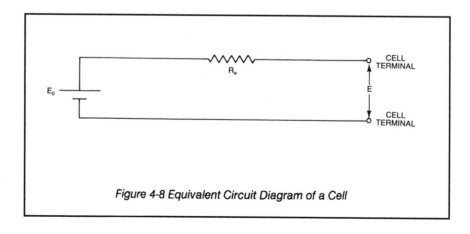

*Figure 4-8 Equivalent Circuit Diagram of a Cell*

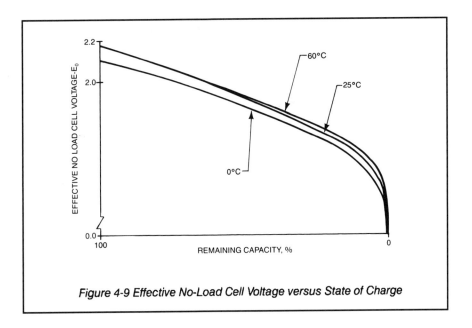

*Figure 4-9 Effective No-Load Cell Voltage versus State of Charge*

into the load and the voltage at the cell terminals (E) is the familiar Thévenin circuit formula:

$$E = E_o - IR_e$$

where:   E = Cell terminal voltage
  $E_o$ = Effective no-load cell voltage
  I = Discharge current
  $R_e$ = Effective internal resistance

### 4.1.4.1   Effective No-Load Cell Voltage, $E_o$

The effective no-load voltage ($E_o$) of a sealed lead cell is a function of the average specific gravity and temperature of the sulfuric acid electrolyte in the cell. When the cell is fully charged, the specific gravity will be at its peak and, correspondingly, the $E_o$ voltage will be at its highest. As the cell discharges, the specific gravity of the electrolyte declines as the acid is gradually converted to water. The no-load voltage of the cell correspondingly decreases as shown in Figure 4-9.

The no-load voltage, $E_o$, discussed in this section differs from the open-circuit voltage of a cell. The effective no-load voltage discussed here is the Thévenin circuit equivalent voltage which is determined by plotting discharge voltage, at a specific state of charge, against discharge rate and extrapolating to zero rate.

### 4.1.4.2   Effective Internal Resistance, $R_e$

The effective internal resistance ($R_e$) is a gross value comprised of a number of smaller contributors which appear in the equivalent circuit analysis as resistive elements. These include the resistivity of the plate grids, the lead posts, and the terminals, and the interface contact resistance between these parts. But, the classic resistive elements represent only a portion of the total $R_e$. Another portion comes from the electrochemical system of the cell including resistance to ionic conduction within the electrolyte, the interface of the electrolyte with the active materials of the plates, and the resistivity of the active materials and their interface with the plate grids. All of these contributors, which when added together make up the total $R_e$ of the cell, will vary independently as a function of changing conditions. The electrochemical

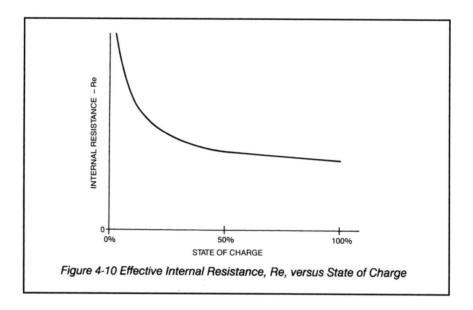

*Figure 4-10 Effective Internal Resistance, Re, versus State of Charge*

components, for example, are affected dramatically by the specific gravity changes of the electrolyte in the cell. The various parameters affecting the gross $R_e$ of the cell are discussed in the following paragraphs. Techniques for measurement of $R_e$ are discussed in Section 4.6.

### 4.1.4.2.1 $R_e$ as a Function of State of Charge

When the cell is fully charged the electrolyte is at its highest state of concentration (highest specific gravity). As the cell discharges, the sulfate ion concentration decreases. This reduction in available current carriers is seen as higher internal resistance in the Thévenin circuit. Figure 4-10 shows this relationship. Notice that no substantial impact occurs until the cell state of charge falls below 25 per cent. Only a gradual increase in $R_e$ occurs from full charge down to 25 per cent state of charge (75 per cent depth of discharge) and then it increases rapidly as capacity approaches zero.

### 4.1.4.2.2 $R_e$ as a Function of Temperature

All the resistive elements (both classically resistive and electrochemical) in the cell are affected by cell temperature. The classically resistive elements vary with temperature on an essentially linear basis. The electrochemistry of the cell has a larger impact on the total $R_e$. This relationship is definitely not linear. At high temperatures the conductivity of the electrolyte is quite good and ionic flow is rapid. As the temperature decreases the conductivity decreases. When the electrolyte approaches its freezing point, its conductivity drops rapidly. Since the freezing point of the electrolyte is a function of the specific gravity, the conductivity of the electrolyte not only depends on temperature but also on the cell state of charge. The effect of temperature on the gross $R_e$, combining both classical and electrochemical elements, is shown in Figure 4-11. The effects of states of charge are also shown.

### 4.1.4.2.3 $R_e$ as a Function of Cell Life

One important effect on the internal resistance of a cell as the cell ages is the increase in the contact resistance between the active material of the plates and the plate grid. The resulting increase in internal resistance is very slight until late in the cell life and then it increases quite rapidly as shown in Figure 4-12.

As the cell is used (charged and discharged), the interface between the current

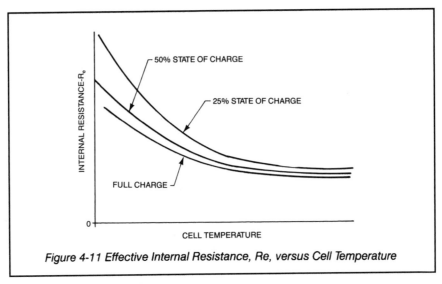

*Figure 4-11 Effective Internal Resistance, Re, versus Cell Temperature*

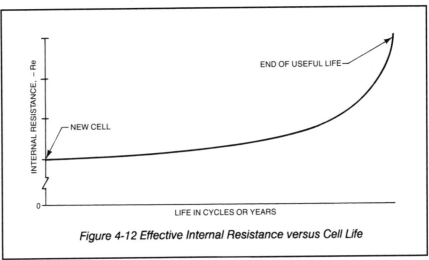

*Figure 4-12 Effective Internal Resistance versus Cell Life*

collector grid and the active material of the positive plate is slowly degraded by oxidation of the grid. The metallic lead of the positive grid is oxidized to $PbO_2$. Since $PbO_2$ is less conductive than the grid, the electrons must flow through an increasingly inefficient current collector as the cell is used. This oxidation process is discussed in detail in Section 4.4. The interface resistance between the grid and the active material also increases. The result is a growth in resistance as the cell ages.

The construction processes and materials used in advanced sealed-lead cells and batteries retard the oxidation process and extend the time and number of cycles before the internal resistance increases significantly.

## 4.1.5 BATTERY VOLTAGE - GENERAL OVERVIEW

In most battery applications, the discharge current is approximately constant and the parameter of concern is the behavior of the battery voltage with time. Constant-power and constant-resistance discharges are also important, but are usually well modeled by a constant-current discharge. So, voltage behavior under various forms of constant-current loads will be the focus of the remainder of the section.

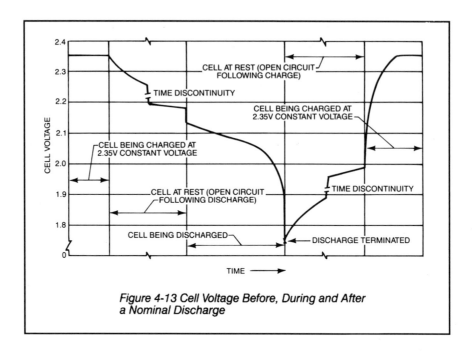

*Figure 4-13 Cell Voltage Before, During and After a Nominal Discharge*

The various stages of a typical battery duty cycle, including charge, discharge, and rest, are illustrated in Figure 4-13.

No matter what type of charger is used, it will hold the battery at some artificially high voltage during the charge process. When the battery is fully charged and removed from the charger, the battery voltage will drop to its full-charge open-circuit value. This value will decay only very slightly as the battery self-discharges.

When the battery is placed on discharge, the voltage will normally drop immediately from its open-circuit value to its on-load value. [For high-rate applications (4C and above), the voltage behavior is somewhat different. See the discussion in Section 4.1.6.] The loaded battery voltage will remain on a plateau, declining only slightly, for most of the battery's useful discharge. When the voltage hits the knee of the curve, the fall to zero volts is extremely rapid. The discharge is normally terminated at this point.

After discharge, if the battery is left at rest in an open circuit condition, the voltage will gradually recover to a level near 2.0 volts depending on the degree of discharge.

### 4.1.5.1 Mid-point Voltage

A common way of evaluating the discharge characteristics of a cell is to use mid-point voltages. Mid-point voltage, by definition, is the voltage of the cell when it has delivered 50 per cent of its capacity at the given discharge rate. In other words, it is the half-way point for any given discharge rate. The voltage characteristic for many sealed-lead batteries (Figure 4-14) is such that the mid-point voltage is also the approximate average voltage for the plateau of the discharge curve. This makes it a convenient point to estimate average performance in terms of voltage delivery to the load. The mid-point voltage concept will be used extensively throughout this section.

### 4.1.5.2 Battery Discharge Voltage as a Function of Discharge Rate

The effects of increased discharge rate on the battery voltage are manifested in three ways: depression of the voltage plateau, an increase in the slope of the plateau, and

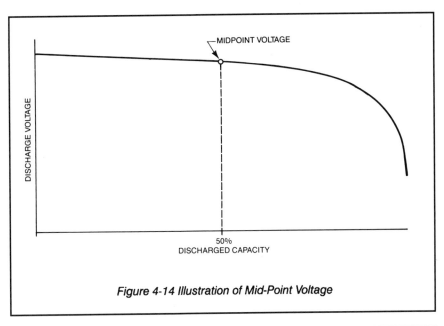

*Figure 4-14 Illustration of Mid-Point Voltage*

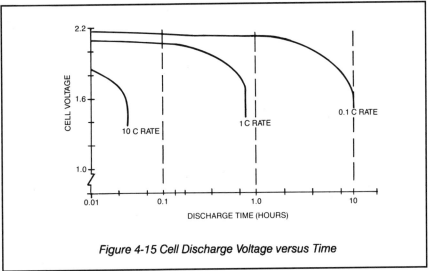

*Figure 4-15 Cell Discharge Voltage versus Time*

shortening of the length of the plateau. Figure 4-15 shows a family of discharge curves for three different discharge rates as a function of time. As can be seen from those plots, low to medium-rate discharges behave similarly. Although there is some voltage depression with the increase in rate, the primary effect is shortening the discharge time. However, the high-rate (10C) discharge behaves quite differently. For this reason, high-rate discharges are discussed separately in Section 4.1.6.

### 4.1.5.3   Battery Discharge Voltage as a Function of Battery Temperature

As the temperature of the sealed lead cell changes, both the $E_o$ and the $R_e$ of the cell vary, impacting the discharge voltage at the terminals of the cell.

The cell's $E_o$ varies slightly with temperature due to minor changes in the interface between the active materials and the electrolyte as well as solution activity effects at differing temperatures. At higher temperatures the $E_o$ is slightly higher and at low temperatures the $E_o$ is slightly lower.

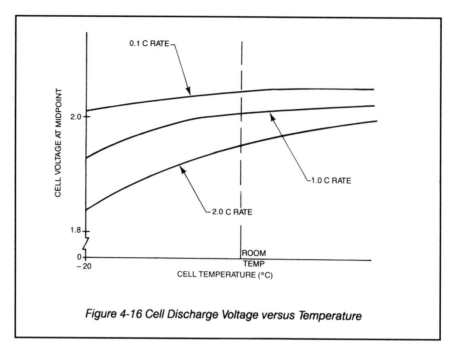

Figure 4-16 Cell Discharge Voltage versus Temperature

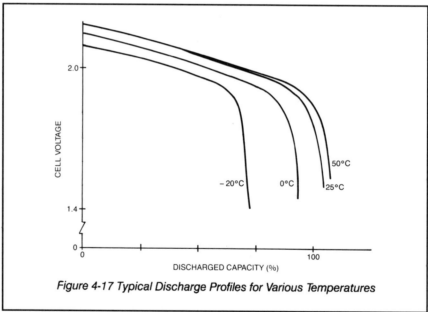

Figure 4-17 Typical Discharge Profiles for Various Temperatures

The impact on cell voltage from changes in $R_e$ with temperature is normally greater than that from $E_o$. $R_e$, as discussed earlier, is a combination of many elements in the cell-resistive and electrochemical. The electrochemical parts of the total $R_e$ vary quite broadly, particularly at low temperatures. The combined temperature effect on $R_e$ was shown in Figure 4-11 to be dramatic. When the temperature effects upon $E_o$ and $R_e$ are taken into consideration, the overall effect can be displayed as a family of curves at different discharge rates as shown in Figure 4-16. Note that this characterizes the discharge profile by its mid-point voltage. An alternative approach to visualizing the effect of temperature upon discharge voltage is shown in Figure 4-17.

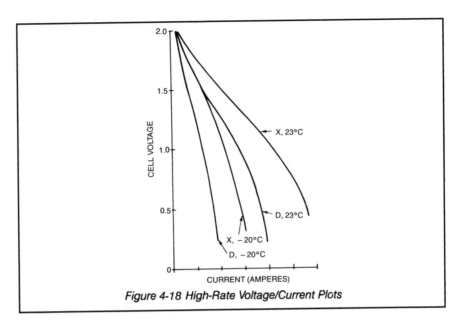

*Figure 4-18 High-Rate Voltage/Current Plots*

## 4.1.6 HIGH-RATE DISCHARGES

The design approach used with many starved-electrolyte sealed-lead batteries make them superlative performers in high-rate applications. This high rate capability has been utilized primarily in engine starting applications and uninterruptible power supplies (UPS). Design of battery systems for high-rate applications is a very specialized field. Some general discussion is provided here, but designers contemplating use of batteries in high-rate applications are encouraged to discuss their needs with battery manufacturers.

Typically in high-rate applications, the principal concern is maximizing instantaneous power through very high currents while still maintaining an acceptable voltage. While the advanced sealed-lead products are capable of delivering very high-rate current for short periods of time, any continuous discharge must be limited to lower currents to avoid damage to the battery. Figure 4-18 shows the relationship between peak current and its corresponding voltage for two sizes of cell at two temperatures. As can be seen, the current at 1.2 volts per cell (a nominal voltage for a high-rate, engine-start application) is over 30C. These E-I traces can also be translated into plots of instantaneous peak power as shown in Figure 4-19. Appendix B includes both E-I and instantaneous power plots.

A constant-current discharge at high-rate has a voltage profile that is significantly different in shape when compared to those shown earlier for lower rate discharges. As can be seen in Figure 4-20, the voltage depression is obviously dramatic and the voltage plateau has a very pronounced slope. In addition, there may be a momentary depression of the voltage below the plateau immediately after the load is imposed. This transient, called *coup de fouet* (whipcrack), is common to all lead-acid batteries discharged at high rates. In most applications, the transient has no effect on the battery system's performance. The areas of concern are systems that have low-voltage cutouts or that have digital electronics that may be affected by a low-voltage transient. Again, this is an area best addressed in consultation with the battery manufacturer.

## 4.1.7 DISCHARGE LIMITS

In order to obtain maximum life from sealed-lead cells and batteries, they should be disconnected from the load once they have discharged their full capacity. In fact,

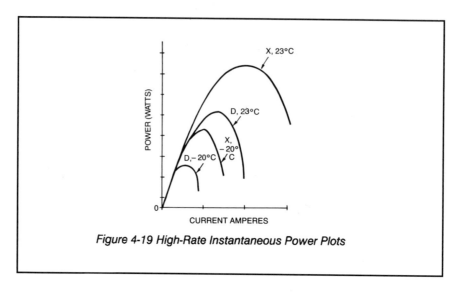

Figure 4-19 *High-Rate Instantaneous Power Plots*

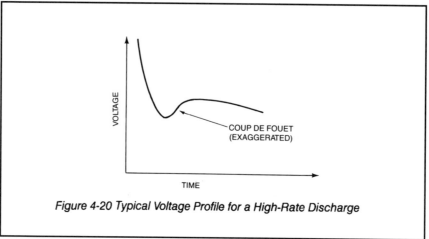

Figure 4-20 *Typical Voltage Profile for a High-Rate Discharge*

once a cell or battery has passed the knee of the discharge curve at the end of the voltage plateau, there is relatively little additional capacity to be extracted from the battery. Disconnecting at that point will minimize the possibility of overdischarge.

In *overdischarge,* the sulfuric acid electrolyte can be depleted of the sulfate ion and become essentially water. This lack of sulfate ions to act as charge conductors will cause the cell impedance to rise and little current will flow. This may necessitate a longer charge time or alteration of charge voltage before normal charging may resume.

A second potential problem arising from overdischarge can occur because of the increased solubility of lead sulfate as the concentration of the acid decreases. In a severe overdischarge condition when the electrolyte has become water, some of the lead sulfate present at the plate surfaces may go into solution. On recharge, the sulfate ion is converted back to sulfuric acid leaving a precipitate of lead metal (dendrite) in the separator. This may then result in a resistive path between the plates causing battery failure.

Because of possible occurrence of the problems described above, keeping the battery connected to the load or allowing it to self-discharge past the point where the battery capacity is depleted is not recommended.

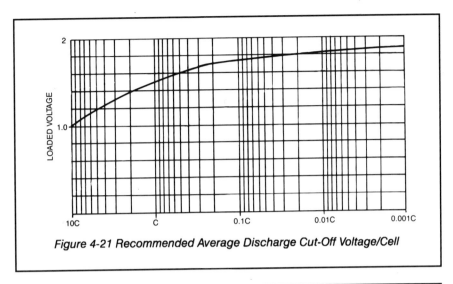

Figure 4-21 Recommended Average Discharge Cut-Off Voltage/Cell

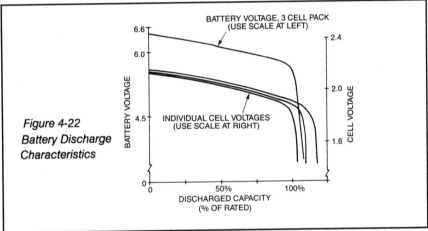

*Figure 4-22 Battery Discharge Characteristics*

### 4.1.7.1 Cell and Battery Discharge Limits

The discharge voltage at which 100 per cent of the usable capacity of the cell has been removed is a function of the discharge rate and is shown in Figure 4-21.

Most battery applications require more than 2 volts, however. This means that cells must be connected in series to make up the required battery voltage. In a series string of cells the battery performance is determined by the behavior of the individual cells. Fortunately, the cells are quite uniform, one to another, in voltage and available capacity. It should be recognized, though, that some variations do exist.

The voltages of the different cells in a given battery are normally very close to each other as the battery is being discharged. The largest impact upon battery voltage, other than an absolute failure, comes from the capacity of the individual cells as the battery is deeply discharged. Figure 4-22 shows how a battery might look as it is deeply discharged. (The battery voltage scale on the left is purposely plotted differently from three times the cell voltage scale on the right to show clearly the deep discharge effect on the battery.)

The battery voltage near the end of useful discharge is determined by the lowest capacity cell in the battery. The knee of the discharge characteristic is sharper than that of the individual cells and once the lowest cell is totally expended, the battery voltage drops rapidly.

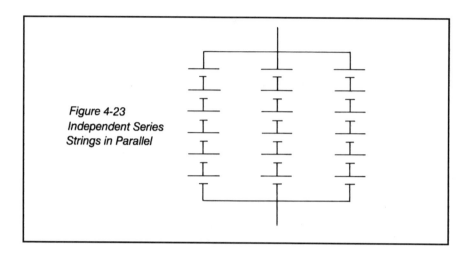

*Figure 4-23*
*Independent Series*
*Strings in Parallel*

### 4.1.7.2   Disconnect Circuits

Leaving the battery connected to a load after discharge should be avoided to enable the battery to provide its full cycle life and charge capabilities. Some form of battery disconnect or kickout circuit is often supplied to remove the battery from the load once the battery capacity is exhausted. After discharge and removal of the load from the battery, the cell voltages will normally increase and stabilize at the open circuit voltage as sketched in Figure 4-13. Because of this phenomenon, some hysteresis must be designed into the battery disconnect circuitry so that the load is not reapplied to the battery under this condition. The disconnect circuit should also be designed so that it does not itself impose a load on the disconnected battery. i.e. the battery is truly open-circuited.

## 4.1.8   DISCHARGING CELLS AND BATTERIES IN PARALLEL

In general, use of a larger battery, rather than multiple batteries in parallel, is both more reliable and more cost-effective. But, situations still arise where batteries need to be connected in parallel.

There are two approaches to connecting cells and batteries in parallel. The most common method shown in Figure 4-23 is the simpler. The method shown in Figure 4-24 is more difficult but may provide increased reliability in some applications. The connection in Figure 4-23 has three packs, each consisting of six cells connected in series, with the packs connected in parallel. The connection in Figure 4-24 has six series connections between groups of three cells in parallel. The difference is important only in deep discharge applications. Some of the cells in any given string in Figure 4-23 could be very deeply discharged before the battery voltage cutoff is reached. There is much less chance of cells deeply discharging before cutoff is reached in Figure 4-24. The arrangement in Figure 4-24 also provides a more even distribution of capacity among all cells during charging. As discussed in Section 4.2.9.2, charging such an interconnected string requires careful use of a constant-potential charger.

The cell-to-cell paralleling of Figure 4-24 is necessary only when the battery is performing one cycle per day or when the discharge period is less than 30 minutes. As the duty cycle becomes easier, the need for cross strapping becomes less. If the discharge period is two hours or less, the batteries need to be interconnected every 6 to 12 volts. If the discharge duty is longer than two hours, interconnections every 24 or 48 volts are sufficient. In deciding whether to interconnect cells, one advantage of

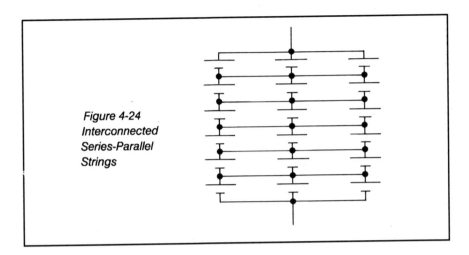

Figure 4-24
Interconnected
Series-Parallel
Strings

the arrangement shown in Figure 4-23 is the ability to remove a battery string while maintaining backup power. This approach is commonly used in many telecommunications applications to allow testing or replacement of individual strings while retaining some backup capacity.

## 4.1.9   SUMMARY

The discharge performance of starved-electrolyte sealed-lead cells and batteries is one of their strengths. Taking advantage of this asset requires understanding the influence that use parameters and environmental factors may have on the resulting discharge. The actual capacity that may be obtained from a battery is greatly affected by the discharge rate and the temperature of the battery. The discharge should be allowed to continue long enough to take advantage of the battery's long voltage plateau on discharge, but the discharge should be terminated before the possibility of overdischarging the battery occurs. Designers using sealed-lead batteries need to be alert to possible overdischarge of the cells, especially when used in batteries comprised of many cells. Precautions such as matching cells by discharge capacity or using special disconnect circuits may be required, particularly in deep-discharge or long-string applications.

# 4.2  Charging

The purpose of any battery is to provide electrical energy upon demand. When the ability of a primary battery's chemicals to deliver energy has been expended, it is discarded. With a secondary battery, such as the sealed-lead battery, the chemical reactions involved are reversible. These batteries may be recharged after discharge so they can deliver their rated output many times over the course of their life.

The type and degree of charging used with a secondary battery is usually a critical factor affecting both discharge performance and life. Insufficient charge input will result in reduced discharge capability. Charging the battery without adequate controls or charging the battery for prolonged periods may reduce its life by any one of several processes.

## 4.2.1  IMPORTANCE OF ADEQUATE CHARGING

All secondary batteries like to be charged fully to give their best performance. Unfortunately, they often can be damaged by being charged too vigorously. When selecting a charging strategy, the need for a full charge must be balanced against the problems associated with overcharging.

In general, experience with sealed-lead cells and batteries indicates that application problems are more likely to be caused by undercharging than by overcharging. Since the starved-electrolyte cell is relatively resistant to damage from overcharge, designers may want to ensure that the batteries are fully charged, even at the expense of some degree of overcharge. Obviously, excessive overcharge, either in magnitude or duration, should still be avoided. Application engineers from the battery manufacturer should be consulted to assist with proper charger design.

## 4.2.2  SEALED-LEAD CHARGING CHARACTERISTICS

From a simplistic viewpoint, charging a sealed-lead battery is analogous to pumping water back into a water reservoir from which it has been removed. But unlike a water reservoir, the battery is not fully charged when the amount of charge returned is equal to that previously removed. There is always some parasitic generation of gas (both oxygen at the positive plate and hydrogen at the negative plate) which reduces charging efficiency. These rates of gas generation are relatively low at low states of charge, but increase as full charge is approached. When the cell is fully charged, essentially all of the charge current is being applied to the generation and recombination of oxygen inside the cell because all of the usable active materials on the plates have been converted to the charged state. The rate at which oxygen recombines at the negative plate is a complex function of cell design, operating conditions, and the overcharge regime. At low rates of charge, the recombination process is efficient, approaching 100 per cent recombination. For a given type of cell or battery, the oxygen recombination efficiency begins to drop at some level of charge current and it continues to decline as the rate of charge increases. The same comments apply generally to hydrogen generation and release. This means that charge currents must be restricted in order to avoid undesirable gas release by the vent as the cell approaches and ultimately achieves a full state of charge. It should be recognized that, although the theory of recombination electrochemistry implies that oxygen generation at the positive electrode will minimize or preclude hydrogen release from the negative, in actual practice, the two processes occur simultaneously and the rates of both are a function of overcharge level.

Conversion of active material at the positive plate is an oxidation process. At full charge, in addition to generation of oxygen, there is also a secondary tendency to oxidize the current-carrying lead grid onto which the positive-plate active material is pasted. This irreversibly converts the metallic lead conductive grid to less conductive lead dioxide. Excessive overcharge current and elevated temperature speed grid oxidation which progressively diminishes the conducting cross section. The ultimate result is conversion of sufficient grid metal to cause loss of electrical continuity between the positive plate active material and the cell terminal. High-purity lead grids minimize the grid oxidation rate.

A third item which must be considered in sealed-lead battery charging is capacity retention. A battery on open-circuit stand will self-discharge. If used in a standby power application, it is not sufficient to fully charge the battery and then leave it on open circuit. The time between discharges may be months or years. At 25°C a fully charged battery will self-discharge to approximately 50 per cent of its rated capacity after one year. Therefore, it is necessary, in these applications, to provide some manner of sustaining charge, normally through a continuous float or trickle charge.

In summary, there are two major reasons for continuing the charging operation into overcharge and likewise two major reasons for maintaining close control over overcharging. With regard to the need for overcharge, full charge is attained asymptotically. This makes it difficult to determine precisely when the battery is fully charged. The battery is thus overcharged to assure that it reaches full charge. Also, overcharge current is maintained to prevent loss of capacity resulting from self-discharge. Control of overcharge current is required to minimize gas venting and to avoid accelerating oxidation of the positive plate grid.

## 4.2.2.1 Cell Pressure, Temperature, Voltage, and Current Interrelationships During Charging

The cell's voltage, current, pressure and temperature are the principal parameters that vary during charging. When those variations are understood, one can then design a charger to effectively and safely charge the cell. The pressure, temperature and voltage profiles are shown in Figure 4-25 for a discharged sealed-lead cell that is charged with a medium charge rate, such as 0.1C constant-current in a 25°C environment.

In Section 2.5.1 the charging chemical reactions were presented. When the cell is at a low state of charge all of the electrical energy input to the cell is converted to chemical energy-producing $PbO_2$ at the positive plate and sponge lead at the negative. During this efficient conversion of the active materials the cell pressure remains low and there is little temperature rise. The voltage rises slowly as the electrolyte is gradually converted from a weak solution to a higher acid concentration. As more electrical energy is pumped into the cell, a gradual change takes place; the positive and negative electrodes can no longer convert all the electrical energy into chemical energy. An increasing amount of charge goes to generate oxygen gas at the positive plate and hydrogen at the negative. This phenomenon is often referred to as *gassing*. Note that gassing is internal to the cell; if the gas escapes the cell, the process is called *venting*.

Cell pressure remains low until the cell approaches 80 per cent state of charge and the positive plate begins to generate oxygen. At the same point or shortly thereafter, the negative plate also begins to gas hydrogen. As the cell gradually transitions from relatively low gassing levels to a condition where the majority of the charge

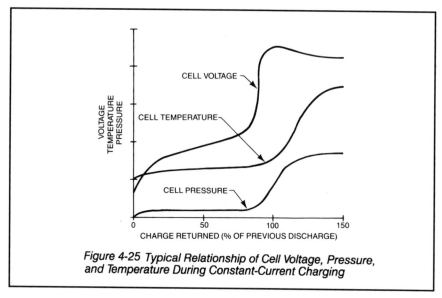

Figure 4-25 *Typical Relationship of Cell Voltage, Pressure, and Temperature During Constant-Current Charging*

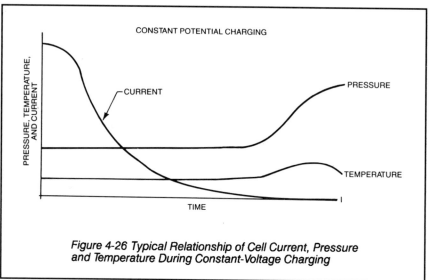

Figure 4-26 *Typical Relationship of Cell Current, Pressure and Temperature During Constant-Voltage Charging*

current is going into gas generation, venting may begin depending upon the over-charge rate. As time progresses the gassing rate and subsequent venting will approach a steady rate that is characteristic of the particular overcharge conditions employed.

The cell temperature profile is similar to the pressure profile, but lags it in time. The cell temperature increases as oxygen recombines at the negative plate releasing heat. The shape of the internal cell temperature curve and the temperature level achieved is again a function of battery design, ambient conditions, and the overcharge regime.

The relatively abrupt increase in cell voltage seen in Figure 4-25 is due to the negative plate going into overcharge with an attendant increase in the rate of hydrogen generation at the negative. The voltage reaches a peak level and then drops off due to oxygen recombination "pulling down" the potential of the negative plate. The shape of this part of the curve is variable, again depending upon a complex set of design and performance parameters.

When a cell is charged by a constant-voltage charger the temperature and pressure relationships are quite different from the constant-current charging situation. Figure 4-26 shows typical parameters for constant-voltage charging. The big difference inside the cell between constant-current charging and constant-voltage charging is the pressure and temperature in overcharge. The charging current drops off significantly in constant-voltage charging as the cell becomes fully charged. The reduction in charging current means that there is less driving force to generate gas and hence less venting. Since there is less oxygen to reduce at the negative there is less heat generated, resulting in only a very small rise in temperature, or possibly a rise and decline as shown, depending on the charge voltage.

## 4.2.3   CHARGE ACCEPTANCE

*Charge acceptance* is the term frequently used to describe the efficiency of charging. If a rechargeable battery were 100 per cent efficient, it would mean that all the energy put into the battery by charging could be retrieved by discharging. But no battery is ideal; no battery is 100 per cent efficient. The charge acceptance of sealed-lead batteries in most situations is quite high, typically greater than 90 per cent. A 90 per cent charge acceptance means that for every amp-hour of charge introduced into the cell, the cell will be able to deliver 0.9 amp-hours to a load. Charge acceptance is affected by a number of factors including cell temperature, charge rate, cell state of charge, the age of the cell and the method of charging. Each of these will be discussed in the following sections.

### 4.2.3.1   Effect of State of Charge on Charge Acceptance

The state of charge of the cell will dictate to some extent the efficiency with which the cell will accept charge. When the cell is fully discharged, the charge acceptance is immediately quite low. As the cell becomes only slightly charged it accepts current more readily and the charge acceptance jumps quickly, approaching 98 per cent in some situations. The charge acceptance stays at a high level until the cell approaches full charge.

As mentioned earlier when the cell becomes more fully charged, some of the electrical energy goes into generating gas which represents a loss in charge acceptance. When the cell is fully charged, essentially all the charging energy goes to generate gas except for the very small current that makes up for the internal losses which otherwise would be manifested as self-discharge. A generalized curve representing these phenomena is shown in Figure 4-27.

### 4.2.3.2   Effect of Temperature on Charge Acceptance

As with most chemical reactions, temperature does have a positive effect upon the charging reactions in the sealed-lead cell. Charging at higher temperatures is more efficient than it is at lower temperatures, all other parameters being equal, as shown in Figure 4-28.

The cell temperature impact on charge acceptance is overlaid upon the generalized curve of Figure 4-27 to illustrate the compound effect of cell state of charge and cell temperature. For this representation the other important parameter, charge current, is held constant. Figure 4-28 shows the charge acceptance is still very high at lower temperatures.

### 4.2.3.3   Effect of Charging Rate on Charge Acceptance

The starved-electrolyte sealed-lead cell charges very efficiently at most charging rates. The cell can accept charge at accelerated rates (up to the **C** rate) as long as the

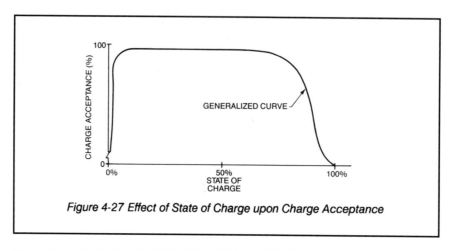

*Figure 4-27 Effect of State of Charge upon Charge Acceptance*

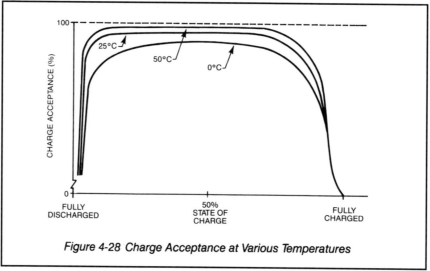

*Figure 4-28 Charge Acceptance at Various Temperatures*

state of charge is not so high that excessive gassing occurs. And the cell can be charged at low rates with excellent charge acceptance.

Figure 4-29 shows the generalized curve of charge acceptance now further defined by charging rates. When examining these curves, one can see that at high states of charge, low charge rates provide better charge acceptance.

### 4.2.3.4    Other Factors Affecting Charge Acceptance

When a sealed-lead cell ages, the ability of the cell to accept charge decreases and the capacity available from each charge is reduced. This is partly due to the reduction in the ability of the positive plate to conduct electrons to the external circuit because of the gradual oxidation of the grid. But it is also due to a lack of continuity in the active material of the plates of the cell, particularly the positive plate, which impacts charge acceptance. For this reason, an aged cell will begin gassing sooner in the charge cycle.

In multi-cell batteries another factor may be involved. The charge acceptance of any given cell in the series string may seem reduced if the battery pack is charged with a constant-voltage charger set at a low voltage. Under this condition, one cell in a battery may have a slightly lower charge acceptance than the other cells. When the battery is discharged, this cell will recover its capacity more slowly than the other

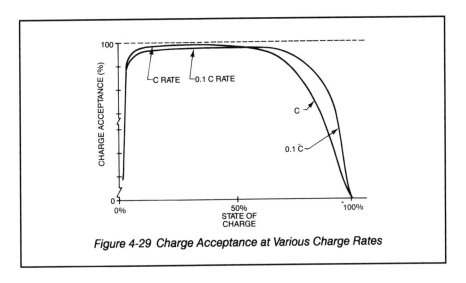

*Figure 4-29 Charge Acceptance at Various Charge Rates*

cells in the pack. It is quite possible that this one cell will not obtain a full charge if the charge time is short before the next discharge. Repetitive rapid cycling of the pack without sufficient recharge will generally cause the low-charge-acceptance cell to become progressively lower in capacity. This problem is often associated with series-parallel applications. In those cases, interconnection between cells as discussed in Section 4.1.8 may reduce the likelihood of cycle-down.

## 4.2.4   OVERCHARGING

*Overcharge* is defined as continued charging of a cell after it has become fully charged. When a cell is not yet fully charged, the electrical energy of the charge current is converted to chemical energy in the cell by the charging reactions. But, when all of the available active material has been converted into the charged state, the energy available in the charging current goes to produce gases from the electrolyte in the cell.

In the starved-electrolyte sealed-lead cell at typical charging rates, the bulk of the gases are recombined and there is virtually no venting of gases from the cell.

### 4.2.4.1   Oxygen Recombination Reaction

As described in Section 2.5.2, when the cell reaches full charge, virtually all of the active material in the positive electrode is charged while the negative electrode, by design, still contains some uncharged active material. This balance of materials results in oxygen being generated at the positive electrode during overcharge prior to generation of hydrogen at the negative. The oxygen that migrates to the negative plate promptly recombines, converting the sponge lead to lead oxide. Since the oxygen recombination reaction is exothermic, heat is liberated as part of the process. The bisulfate form of $HSO_4$ in the sulfuric acid electrolyte then reacts quickly with the lead oxide to form lead sulfate with water as a byproduct. This lead sulfate at the negative plate is then reconverted back to sponge lead and sulfuric acid by the overcharge current. Thus, although oxygen is being generated at the positive plate during overcharge by the breakdown of water, it is being converted, or *recombined* back to water by the chemical reactions occurring at the negative plate.

Although the recombination process theoretically keeps the negative plate from going into overcharge by continuously forming lead sulfate by reaction of oxygen with the sponge lead, under many operating conditions some hydrogen is generated at the negative in overcharge. In principle this hydrogen can migrate to the positive electrode and chemically recombine back to water by a series of reactions similar to those described for oxygen recombination, but this is not known to occur readily. Instead small amounts of hydrogen escape from the cell under some operating conditions, depending on the level of overcharge.

These reactions continue as long as overcharge current flows. However, little electrolyte is lost because venting is minimal under most circumstances due to the recombination process. The vast majority of the electrical energy in this equilibrium overcharge process is converted to heat energy raising the temperature of the cell. The oxygen recombination capability of the starved-electrolyte cell permits true maintenance-free operation. While the recombination process is highly efficient under most normal charging conditions, small amounts of hydrogen and/or oxygen may be vented from the cell. However, the amounts of these gases are much lower than those discharged by other types of lead batteries being charged under equivalent conditions. There are, of course, situations of high overcharge currents and/or abnormal temperatures that may result in excessive levels of gas venting from the cell. These conditions, if allowed to persist, may permanently damage the cell.

### 4.2.4.2 Oxidation Effects of Overcharge

One other aspect of overcharge may also be detrimental to the life of the sealed-lead cell. Even though no electrolyte may be lost in the overcharge reactions described above, and even though the inorganic separator is not degraded by the temperature and oxygen created by the overcharge reactions, it is still best to limit overcharge to minimize oxidation of the positive grid.

The positive plate, made up of a metallic lead grid onto which the active material is pressed, undergoes a gradual change during overcharge. The grid metal, when exposed to sulfuric acid and elevated temperature in overcharge, oxidizes forming lead dioxide. This reaction takes place quite slowly because the grid is completely surrounded by active material ($PbO_2$) which tends to isolate or shield the grid from the electrolyte. As the grid is gradually oxidized, the lead cross-section is reduced and the current-carrying capability of the grid is correspondingly reduced because lead dioxide is a relatively poor conductor. Eventually the oxidation may become so severe that the grid ceases to adequately perform the function of conducting electrons to the external circuit. Oxidation of the grid also creates other problems. Lead dioxide occupies a greater volume than the lead it replaces. As the positive grid oxidizes, the positive electrode thus expands. The resulting plate "growth" can result in electrical shorts or physical damage, such as buckling, to the positive plate. The cell also becomes increasingly fragile as the structurally weaker lead dioxide replaces the metallic lead in the grid. The high-purity lead grids found in advanced sealed-lead batteries minimize the oxidation rate thus delaying the onset of these problems.

The oxidation of the lead grid can be delayed, thus extending the life of the cell, by minimizing the amount of overcharge. Both the magnitude and the duration of the overcharge have a direct influence upon the rate and degree of oxidation of the grid.

Again, it should be stressed that the majority of application problems with sealed-lead batteries are attributable to undercharging. Many power supply designers may be willing to accept a slight decrease in cell life as the price for increased assurance of meeting the application requirements.

### 4.2.4.3 Overcharge Characteristic - The Tafel Curves

When a cell is in overcharge in a state of equilibrium, the voltage/current relationship is quite predictable. The cell voltage versus overcharge current characteristics, called the Tafel curve, can be used by the charger designer to select the proper con-

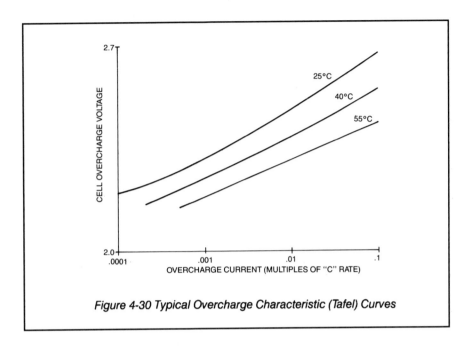

*Figure 4-30 Typical Overcharge Characteristic (Tafel) Curves*

stant voltage setting for a specific charger/ battery application. Typical Tafel curves for sealed-lead cells are shown qualitatively in Figure 4-30 and repeated with the appropriate scales in Appendix B. The overcharge voltage for any given current is quite dependent upon temperature, but as can be seen, it is not a linear relationship. A typical sealed-lead cell will have an overcharge voltage that is inversely related to temperature.

## 4.2.5  TYPES OF CHARGING

Charging the sealed-lead battery, like charging other secondary batteries, is a matter of replacing the energy depleted during discharge.

Charging may be accomplished by various methods, but the objective — to drive current through the battery in the direction opposite that of discharge — remains the same. Constant-voltage charging is conventionally used for lead-acid batteries and is fully acceptable for the starved-electrolyte sealed-lead battery; however constant-current, taper-current and variations thereof may also be used. Each method has its advantages and disadvantages which will be discussed in the following sections. The choice of charging approach should be determined by its fit with the application's requirements.

When considering the charger to use, it is necessary to consider the way in which the battery will be discharged, the time available for charge, the temperature extremes the battery will experience, and the number of cells in the battery.

As described in Section 2.2.4, cyclic applications are those where the battery is frequently discharged, removed from the load and subsequently charged. For this discharge profile, typified by portable instrumentation and battery-powered appliances, constant-voltage, constant-current or two-step constant-current chargers may be used. The essential requirement for chargers in this type of duty is the ability to restore full charge quickly.

In float applications, the battery is continually charged, discharging only when the main power has failed. In operations of this type, which include memory back-up, emergency lighting and alarms, uninterruptible power supplies (UPS) and telecommunications systems, a constant-voltage charger normally represents the most effective alternative. The key requirement is to keep overcharge effects at a minimum.

Applications in which the higher cost of a well-regulated constant-voltage or constant-current charger significantly affects the total product cost may require a taper-current charger. Although their low cost is a definite advantage, their lack of voltage regulation can be detrimental to the life of the battery. Taper chargers are often seen in low-cost consumer items, such as rechargeable flashlights.

Charging the sealed-lead battery in remote locations may be accomplished by photovoltaic cells on solar panels supplying energy to charge the battery. In this application a constant-voltage output is preferred to minimize the effects of light intensity and temperature variations.

All of these methods of charging the starved-electrolyte sealed-lead battery are discussed in the following paragraphs.

## 4.2.6 CONSTANT-VOLTAGE (CONSTANT-POTENTIAL) CHARGING

Constant-voltage charging, often referred to as constant-potential (CP) or float charging is the most common method of charging lead batteries. It has been used successfully for over 50 years with a diversity of lead battery types.

Constant-voltage charging means simply that the charger voltage is held uniform regardless of battery state of charge. The charge current varies depending on the difference between the input voltage and the battery voltage: when the battery is discharged, its voltage is lower and the charge flow is greater; as the battery charges, its voltage increases and the charge current declines. Many advanced forms of sealed-lead batteries do not require a current clamp or other limitation on inrush current thus simplifying charger design.

### 4.2.6.1 Uses of Constant-Voltage Charging

Constant-voltage chargers are most often used in two very different modes: as a fast charger to restore a high percentage of charge in a short time or as a float charger to minimize the effects of overcharge on batteries having infrequent discharges, i.e. in most standby power applications. In between these two extremes is an area of cycling-type duty where the constant-voltage charger is not as well-suited. These rather contradictory statements can be explained by remembering that supplying electrons is the key to charging batteries. Because sealed-lead batteries have very low internal resistances, they will accept very high currents when fully discharged if not limited by the current-supply capabilities of the voltage source. Initial charge currents as high as 4C are common on large power supplies. Although the current on a constant-voltage charger starts out extremely high, as illustrated in Figure 4-31, it then decays nearly exponentially.

The important parameter is the area under the curve, the integrated product of current supplied over time. Thus, in the early going, with large current flows, a constant-voltage charger may return as much as 70 per cent of the previous discharge in the first 30 minutes. This proves useful in many applications involving multiple discharge scenarios. As the battery charges, its voltage increases quickly, reducing the potential difference that has been driving the current, with a corresponding rapid decay in the charge current. As a result, even though the battery quickly reaches partial charge, obtaining a full charge requires prolonged charging. The length of the charge period may be significantly affected by the choice of charge voltage.

Given this behavior, constant-voltage chargers are frequently found in applications that normally allow extended charging periods to attain full charge. Here the

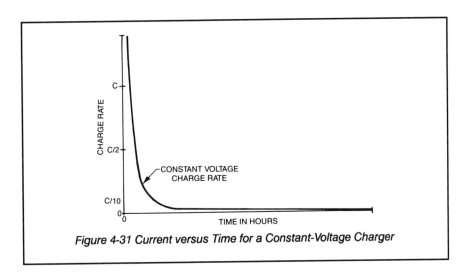

*Figure 4-31 Current versus Time for a Constant-Voltage Charger*

fast charging to a significant fraction of the previous discharge is a significant advantage in accommodating multiple power-loss situations. Exploiting the constant-voltage charger's fast charging features in a cycling application is difficult, since repeated discharges without the intervening time to asymptotically reach 100 per cent capacity, will result in cycling down of the battery.

### 4.2.6.1.1 Cycling Down and Charge Time

It is often very tempting to trim the charge time for batteries on constant-voltage charge. After all, the battery has quickly regained the vast majority of the capacity used on the previous discharge and charge currents are nearly negligible. This approach may work occasionally, but it is a practice which, if repeated often, will result in diminished battery capacity.

As explained in section 4.2.7.2, repeated discharges, especially deep discharges, and constant-voltage charges may cause the battery to become imbalanced which can then lead to failure. The problem is that a low-capacity cell, if it is present, is more heavily drained than the remainder of the battery. Consequently, even though the other cells in the battery are charged to capacity and the charge current has diminished significantly, the low-capacity cell still needs more charging. If the time between discharges is long enough, the low cell can charge as well and the battery will approach balance. But, if that time is not allowed and, instead, the battery is discharged again, the low cell will be even more heavily discharged. Thus the problem is compounded. The term *cycling down* is used to describe this cumulative loss of capacity. The result can quickly become a battery with one or more dead cells. This can be prevented by either adjusting charge voltages or charge times to ensure adequate charging for all cells.

### 4.2.6.2   Approaches to Constant-Voltage Charging

Described below are typical ways of applying constant-voltage charging to sealed-lead batteries.

### 4.2.6.2.1 Single-Potential Charging

The most common sealed-lead battery charger is the single-potential constant-voltage charger which is set for a specified voltage. These chargers normally include some sort of series regulation or shunt regulation between the charging power source and the battery to hold the voltage across the battery terminals relatively constant.

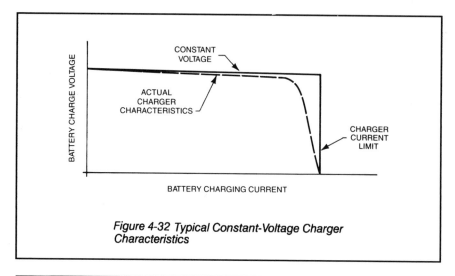

Figure 4-32 *Typical Constant-Voltage Charger Characteristics*

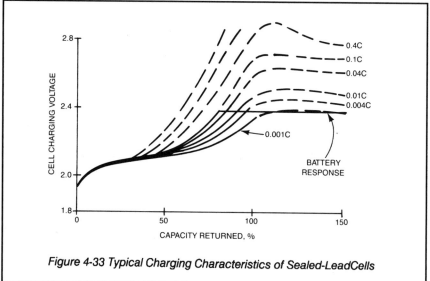

Figure 4-33 *Typical Charging Characteristics of Sealed-LeadCells*

Even so, in practice, the "constant" voltage is not perfectly constant throughout the range of charging currents, as shown by the dotted line in Figure 4-32. For most applications, the actual charger profile is well approximated by the idealized charger curve shown as the solid line in Figure 4-32. The idealized curve assumes the charger will supply a uniform specified voltage at any current below the current limit and then will supply only that current regardless of voltage.

The charger then must interact with the characteristics of the battery. Figure 4-33 represents a typical battery response to charging at various charge currents as the family of profiles shown as the dotted lines. The solid line represents the behavior of the battery when charged by the idealized constant-potential charger of Figure 4-32. The battery first behaves like it is receiving a constant-current charge at the limiting current until the charging voltage reaches the setpoint voltage of the charger. At that point, the battery charging voltage remains constant as the battery continues to charge, i.e. the battery moves to the right on Figure 4-33, cutting the charge current profiles. As can be seen on the figure, as the battery approaches full charge the currents decline quickly. This trend continues into overcharge until, at some point, the

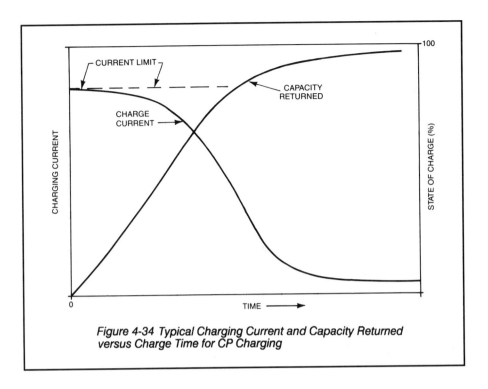

*Figure 4-34 Typical Charging Current and Capacity Returned versus Charge Time for CP Charging*

current reaches a level where it just balances the losses within the cell. This stable trickle current will then continue indefinitely.

Since the capacity returned is simply the integration of current over time, it is possible to relate the battery behavior to charge time as shown in Figure 4-34. The dramatic reduction in charge current as the cell approaches 100 per cent of the capacity returned is well illustrated. This decrease in current is reflected in the capacity returned. The capacity returned, which was linear and increasing rapidly as long as the charger was operating current-limited, takes an exceptionally long time to actually reach 100 per cent. Since there are losses in the charging process, return of about 120 per cent of discharged capacity is often recommended.

Figures 4-34 and 4-35 illustrate the dilemma in selecting a charge voltage for constant-potential charging. For faster charging, the voltage should be set higher, but the result is then a larger trickle-charge current which may degrade cell life.

### 4.2.6.2.2 *Dual-Potential Constant-Voltage Charging*

In some cases a two-step charger may provide better performance than a single potential setting for a constant-voltage charger. The concept is used to avoid the dilemma discussed above. At the beginning of charge a high voltage setting produces a rapid charge, and then, once the battery is fully charged, the voltage is switched to a conservative float voltage to enhance life. Considerable testing with this concept has demonstrated insufficient improvement over the basic single-potential constant-voltage charger to justify the extra complexity in most standby power situations.

### 4.2.6.3  **Charging Parameters**

The choice of charging parameters (charge voltage and current limit) requires weighing economics, battery life, and operational considerations. Increased battery temperature, a higher current limit on the charger, or elevated charger voltage setting all serve to decrease battery charge time.

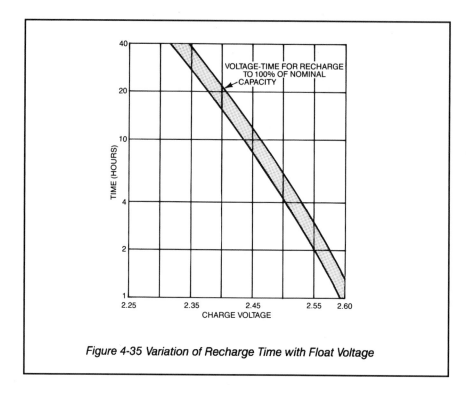

Figure 4-35 Variation of Recharge Time with Float Voltage

### 4.2.6.3.1 *Float Voltage*

The float voltage setting has a dramatic impact upon charge time as can be seen in Figure 4-35. Small increases in float voltage significantly reduce the charge time. Unfortunately, the float voltage setting also has a significant impact on battery life. Since higher float voltage settings permit higher overcharge currents to flow from the charger to the battery, small increases in float voltage can substantially decrease battery life.

To achieve the optimum life, it is best to design for an overcharge rate of approximately .001C at room temperature. By using the Tafel curves, Figure 4-30 or Appendix B, the charger designer can set the float voltage to meet these limits for the specific temperature range in the application.

### 4.2.6.3.2 *Current Limits*

The starved-electrolyte sealed-lead battery is not intrinsically limited in the charge currents it will accept. It is perfectly acceptable to use chargers rated at 20C or more to charge the battery. Practically, greater current-delivery capability in a CP charger normally comes at greater cost. The result is a desire to use relatively small, current-limited chargers, even at the penalty of increased charging time. The impact of various current limits for one typical case are shown in Figure 4-36. A useful rule of thumb for chargers that are limited to less than the 2C rate is to increase the charge time in hours by the reciprocal of the current limit. Thus a charger limited to 0.2C would have 5 hours added to its charge time while a charger with a 0.1C limit would have its charge time increased by 10 hours.

### 4.2.6.4 **Temperature Compensation of Constant-Voltage Charging**

A large improvement in the constant-voltage charger can be made through compensating the voltage setting for changes in temperature. In Section 4.2.3.2 the voltage dependence upon temperature in a sealed-lead cell was discussed. The range of vari-

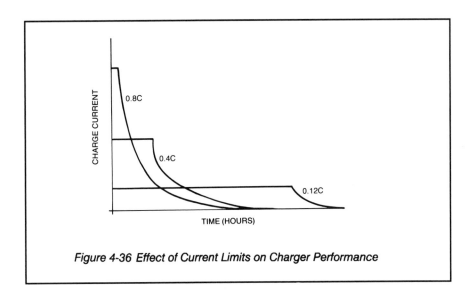

*Figure 4-36 Effect of Current Limits on Charger Performance*

ation in voltage with temperature is a few millivolts per degree C depending upon cell temperature and overcharge rate. Designing the charger to vary the float voltage based on the actual battery temperature will produce a more reliable and long-life charger/battery system. This is obviously most important for batteries routinely operated at other than standard room temperature conditions. It is especially consequential for batteries that spend much of their life in electronics enclosures where temperatures can climb substantially above ambient conditions.

The recommended amount of compensation is provided in Appendix B as a function of charge rate and temperature.

#### 4.2.6.5   **Tradeoffs in Constant-Voltage Charger Design**

In many circumstances, the product designer must balance the trade-offs of faster charge time with higher float voltage settings on one hand versus longer total battery life with lower float voltage settings on the other hand. Some general recommended values are provided in Appendix B. For specific suggestions on the proper charging voltage for a given application, contact the battery manufacturer.

### 4.2.7   **CONSTANT-CURRENT CHARGING**

Starved-electrolyte sealed-lead batteries may be charged with a constant-current charger. In many applications, it is the best method of restoring charge relatively quickly without adversely affecting battery life. Constant-current charging works to eliminate any charge imbalance in the battery and is especially effective when several cells or batteries are charged in series.

Constant-current charging in the practical sense means simply that the charger furnishes a relatively uniform current, regardless of the battery state of charge or temperature. These chargers are inexpensive and, when properly designed, are very reliable. Typically, the charger characteristic is designed so that at the operating point only small variations in current can occur throughout the entire charging range of the battery. Figure 4-37 demonstrates this concept with a theoretical charger voltage-current output characteristic. Superimposed upon this charger characteristic is the operating range of a battery during charging. The difference between maximum current and minimum current in this example is 105mA (max) to 75mA (min) or a range of 30mA. This obviously is not constant-current, but it is acceptable for most

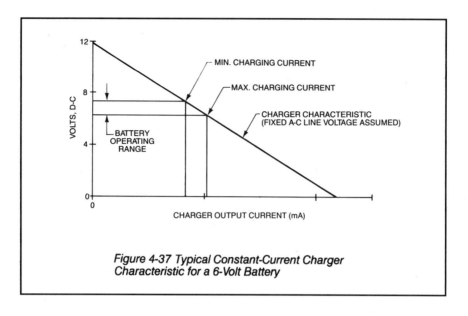

**Figure 4-37 Typical Constant-Current Charger Characteristic for a 6-Volt Battery**

applications. As a sealed cell is being charged with a constant-current charger, the cell charging voltage takes on a profile with time represented by Figure 4-38. This family of curves has been normalized by the percentage of capacity returned to the cell (rather than charging time) so that the various charge rates can easily be compared. As shown by these curves, the voltage of the cell increases sharply as the full charge state is approached. This increase in voltage is caused by the plates going into overcharge. The voltage increase will occur at lower states of charge when the cell is being charged at higher rates due to reduced charging efficiency at higher charging rates. Note that on the far right side of the figure, the charging voltage stabilizes in overcharge at the levels shown in the Tafel curves, Figure 4-30. The family of curves in Figure 4-38 are idealized in that they assume the cells are fully stabilized. In fact, newly manufactured cells may not have such a high peak voltage value as shown. After these new cells have been exercised by discharge/charge cycling, the voltage peak will increase to its stabilized value.

If we now refer back to the charger characteristic in Figure 4-37 which has a nominal charging current of 0.04C and pick off from Figure 4-38 the maximum and minimum voltages of the cell during charge at 25°C, we can predict accurately the actual maximum and minimum charging currents. At other temperatures the charging voltage characteristic will be slightly different than what is shown in Figure 4-38. Rather than plot a whole series of charging voltage characteristics at different temperatures, it is easier to refer back to the Tafel curves (Figure 4-30) to see what happens in overcharge.

It is the overcharging of the sealed-lead battery which is of concern in constant-current charging. Section 4.2.4.2 pointed out that persistent overcharging of a sealed-lead cell increases the oxidation of the positive grid. Therefore, charge limitation becomes a consideration as it relates to battery life when constant-current charging is employed. The next few sub-sections will deal with this issue.

### 4.2.7.1    Single-Rate Constant-Current Charging

Most constant-current chargers are simple single-rate chargers. They are readily available from charger manufacturers or they can easily be designed into the end product. Single-rate constant-current chargers are frequently chosen for portable cyclic applications such as garden tools, flashlights and portable appliances. The

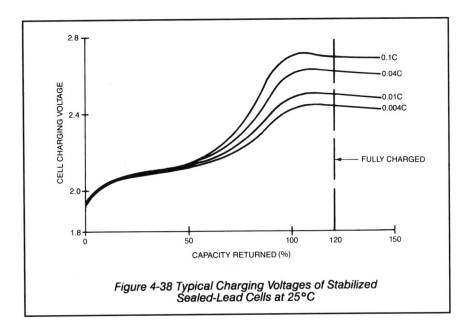

**Figure 4-38 *Typical Charging Voltages of Stabilized Sealed-Lead Cells at 25°C***

prime advantage of constant-current chargers is their low cost. Single-rate constant-current chargers require no switching or sensing circuits and generally consist of a simple transformer and diode rectifiers.

### 4.2.7.1.1 *Charge Rate Selection*

The selection of the proper charge rate is a very important consideration. Choice of charge rate is a function of both the product and the application profile.

To help the product designer in this decision, the maximum charge rate capability must be understood. The starved-electrolyte sealed-lead battery is capable of being charged at very high rates as long as the state of charge is low. Charge rates of up to 1C rate are tolerable for fully-discharged batteries. But as the cell approaches full charge, gassing begins and the internal pressure starts to build. At this point the charge rate must be reduced so that venting from excessive gassing is minimized. For single-rate chargers, then, high charge rates (greater than 1C) are not recommended.

At low temperatures there are two concerns that act to limit the maximum charging rate that may be used. First, at low states of charge and very low temperatures the concentration of the electrolyte solution is so low that the freezing point may be reached. The battery should not be charged when the temperature is below the freezing point of the electrolyte. At temperatures just above the freezing point the ionic conduction through the electrolyte is so sluggish that the charge rate should be reduced well below the room temperature rate. The second concern at very low temperatures is at high states of charges as the cell is approaching and going into overcharge. The gases generated must be minimized to hold the internal gas pressure to a reasonably low level. Since charging at low temperatures is less efficient, gassing may start earlier which again suggests lower charge rates are appropriate.

The other end of the spectrum is the minimum charge rate that can be applied to a battery. As pointed out in Section 4.2.3.3, the charge acceptance of the sealed-lead cell is very good at low charge rates. At the .01C rate, for example, the charge acceptance throughout most of the charging period is approximately 90 per cent at 25°C. The advantage of a very low charge rate is extended life, but this advantage is frequently discounted when it is realized that the charge time will exceed 100 hours for a fully discharged battery.

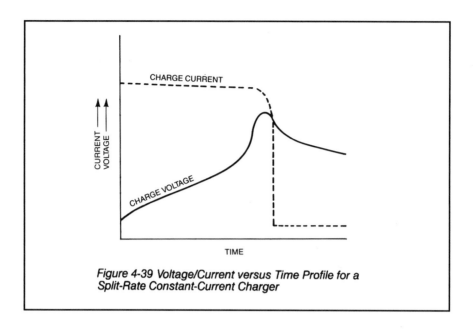

*Figure 4-39 Voltage/Current versus Time Profile for a Split-Rate Constant-Current Charger*

Single-rate constant-current charging is more appropriate for cyclic operation where a battery is often required to attain a full charge overnight or within 24 hours. This suggests a charge rate in the .05C to .1C range. At these charge rates, there will be some venting of gases and positive grid oxidation may diminish life at elevated temperatures and/or extended overcharge times. Normally the user in a cyclic application is instructed to remove the battery from a single-rate constant-current charge within a period of time that is adequate to permit full charge, yet results in minimal grid oxidation.

## 4.2.7.2    Split-Rate Constant-Current Charging

The concept of split-rate constant-current charging is accepted as one of the best methods of charging sealed-lead batteries. The idea is very simple. For the first part of the charging period, while the battery is in a low state of charge, the charge current will be a medium to high charge rate. At a certain point, usually when the battery is approaching full charge, the charging current is switched back to a very low trickle charge rate. The charger then may be left connected to the battery indefinitely.

### 4.2.7.2.1 Advantages of Split-Rate Chargers

Figure 4-39 illustrates the voltage/ current versus time profile for one form of a split-rate constant-current charger based on a combination of voltage and time. Note how the charger switches from the high rate to the low rate when the battery is approximately 110 per cent charged. The choice of switching method and switch point may be affected by the relative priority of minimizing venting (by early switching) versus maintaining good cell balance (by later switching). In some forms of split-rate chargers, during the time when the battery is between 90 per cent and 120 per cent charged, the charger will alternate between the high rate and the low rate, with the amount of time spent in low rate increasing as the battery approaches full charge.

Chargers of this type are extremely useful where the battery charge and discharge pattern cannot be simply described as float or cyclic, but varies between the two applications. Many products, such as portable instrumentation, can benefit from the advantages of split-rate charging. For these uses, the benefits of improved battery performance and life outweigh the increased cost and complexity of the split-rate charger.

The advantages of split-rate constant-current charging are:

- The battery can be charged very rapidly.
- The battery will receive a very limited amount of overcharge because the trickle charge rate is so low that the oxidation effects on the positive plate are minimal.
- The cells in a long series string can be kept reasonably well balanced, one cell to another, throughout the life of the battery.

Each of these advantages is amplified below:

Depending on the expected temperature range of a given application, a high-rate charge current can be selected to fall within the maximum recommended charge rate curves shown in the Appendix B thereby allowing the use of high charge rates during the initial portion of the charge period.

Once the battery reaches a certain point, the charging current is switched from the high rate to a trickle-charge rate. Use of a two-step constant-current charger for a battery in a standby power application ensures that the low rate current is enough to keep the battery charged yet not so high as to create excessive oxidation of the positive grid. In constant-voltage chargers, by contrast, the float current is subject to variations as the battery ages or its temperature fluctuates.

The third advantage of constant-current split-rate charging is better cell balancing than in a constant-voltage charger. If a standby power or float application requires a high voltage, exceeding 40 volts, charging by constant-current means is usually the best method. Consider the example of a 48 volt battery with 24 cells connected in series. Statistically, one or more of the 24 cells will have an inherently lower capacity than the others in the pack. The low-capacity cell may be lower initially, or as the cells age, its capacity may degrade more rapidly that the remainder of the string. As the battery is used in service, after each discharge the low capacity cell becomes lower in relative capacity compared to the rest of the string if the battery is charged with a constant-voltage charger set at a typical float voltage. Assume that there are 23 cells that all have exactly the same capacity and one cell that has a capacity that is 10 per cent lower. And further assume that the battery pack is discharged down to 1.80 volts/cell before charge. The battery voltage at 1.80 volts/cell is 43.2 volts (1.80 x 24 cells). At the point where the entire battery reads 43.2 volts, 23 cells would average approximately 1.87 volts and the low cell would be down to about 0.19 volts. (It is conceivable that the low cell could actually have been driven into reverse.) In any event, the one lower capacity cell has been discharged far more deeply than the other cells in the battery.

Remember from Figure 4-27 that the charge acceptance varies greatly with the cell's state of charge and that deeply discharged cells are initially not very efficient in accepting charge. So the low cell's charge acceptance is significantly less than the other 23 cells in the pack. Not only did the low cell start out 10 per cent behind the others in capacity, it also is not being charged as well as the others due to its initial lower charge acceptance. If the charging voltage were set at 2.35 volts per cell in this example, the total battery float voltage would be 56.4 volts. The 23 higher cells would approach the float setting sooner than the low capacity cell even though the low capacity cell needs additional charge input to catch up. For this example, at the float point, where the current has dropped significantly, the 23 high capacity cells might typically be at about 2.36 volts/cell while the one low cell would still be down at 2.12 volts. Only if the battery were allowed to float for many hundreds of hours before the next discharge, would the low capacity cell become fully charged. And, if

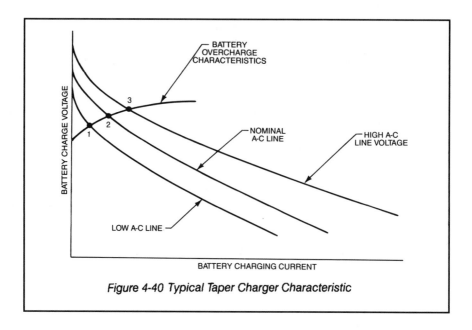

*Figure 4-40 Typical Taper Charger Characteristic*

the next discharge comes before the low cell has caught up, the discrepancies will only amplify. This shows a shortcoming of constant-voltage charging for large multi-cell batteries (more than 20 cells) which can be overcome with a well-designed split-rate constant-current charger. As illustrated in Figure 4-39, the high-rate charging current can, by use of a timer, be continued for a short duration into overcharge to allow all the cells to reach a full charge status before cutting back to the trickle charge rate.

## 4.2.8    TAPER CHARGING

Taper chargers are frequently used on sealed-lead batteries for portable cyclic applications. The taper charger is an unregulated constant-voltage charger. It is designed so that the battery, at low states of charge, receives a high charge rate; as the battery approaches full charge the charge rate decreases. The charge rate in overcharge is still quite high—too high for a standby power application and for many cyclic duty requirements. Figure 4-40 illustrates a typical charger characteristic.

The output current of a taper charger is usually quite susceptible to variations in AC line voltages. Three conditions are plotted in Figure 4-40: the nominal line voltage, a high line voltage, and a low line voltage condition. Superimposed on the charger characteristic is the overcharge characteristic of the battery which is the Tafel curve (Figure 4-30) replotted on this voltage-current graph.

The nominal overcharge current is shown on this generalized curve at point 2. If the AC line voltage is at its very lowest level, the overcharge current would drop to point 1. If a very high line voltage condition exists, the overcharge current increases to point 3. These variations in overcharge current with AC line conditions are very much the same in principle, but greater than, the variations associated with the unregulated constant-current charger described earlier in Section 4.2.7. The overcharge current, even at point 1, the low AC line condition, is too high to be considered acceptable for long term charging. For this reason the taper charger is generally categorized with constant-current chargers. The discussion presented in Section 4.2.7 on constant-current chargers also applies to taper chargers.

A taper current charger contains a transformer for voltage reduction and a half or full-wave rectifier for converting from AC to DC. The output characteristics are such that as the voltage of the battery increases during charge, the charging current decreases. This effect is achieved by the proper wire size and the turns ratio. The turns ratio from primary to secondary determines the voltage output at no load and the wire size in the secondary determines the current at a given voltage. The transformer is essentially a constant-voltage transformer which depends entirely on the AC line voltage regulation for its output voltage regulation. Because of this method of voltage regulation, any changes in the input line voltage directly affect the output of the charger. Depending on the charger design, the output-to-input voltage change can be more than a direct ratio, e.g., a 10 per cent line voltage change may produce a 13 per cent output voltage change.

There are two major charging requirements which must be met if the main concern is the charge time to 100 per cent nominal capacity for cycling applications. These parameters can be defined as the charge rate available to the cell when the cell is at 2.2 volts and 2.5 volts. As shown in Figure 4-40, 2.2 volts represents the charge voltage at which approximately 50 per cent of the charge has been returned to the cell at nominal charge rates of 0.1C to 0.02C. The 2.5 volt point represents the voltage at which the cell is in overcharge. Given the charge rate at 2.2 volts, the charge time for a taper charger can be defined by:

$$\text{Charge time} = (1.2 \times \text{Capacity discharged previously})/(\text{Charge rate @ 2.2V})$$

It is recommended that the charge rate at 2.5 volts be kept between 0.05C maximum and 0.01C minimum thereby ensuring the battery will be charged at normal rates while keeping the battery from being severely overcharged if the charger is left connected for extended time periods.

## 4.2.9    SPECIAL CHARGING SITUATIONS

In addition to the "normal" charging processes discussed in the previous sections, there are certain situations where extra care or a different approach to charging is indicated. Some of these are discussed below.

### 4.2.9.1    High-Voltage Batteries

Charging of high-voltage (over 40 volts) batteries is a concern for two major reasons: personnel safety and charge balancing. Although both items represent areas for concern, they can both be accommodated by the use of appropriate design measures and applications precautions. Batteries with voltages in the 40 to 60 volt range have been used quite successfully in a variety of applications. Batteries with higher voltages have been constructed and used successfully, but both safety issues and application methods require much more attention.

#### 4.2.9.1.1 Personnel Safety

Underwriter's Laboratories recognizes 42 volts as the point at which safety becomes an issue. Charging systems using voltages over this limit should ensure that the charging circuitry is shielded and protected, that batteries are well-insulated, and that personnel access to batteries and energized charging systems is carefully controlled.

#### 4.2.9.1.2 Charge Balancing

The problems of cycling-down and charge-imbalance have been discussed previously. These issues become more important as the number of cells in the battery increases. In general, a good solution is two-step constant-current charging to ensure proper charge. However, many batteries in the higher-voltage regime are "float charged" by direct connection to a rectifier power supply and/or are connected in parallel for

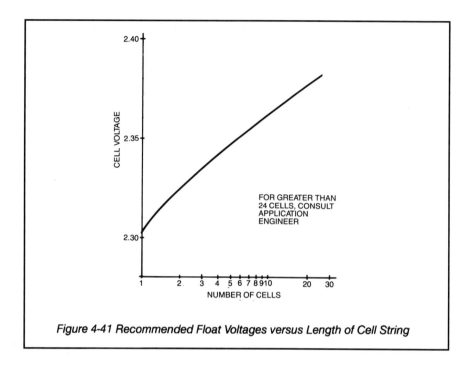

*Figure 4-41 Recommended Float Voltages versus Length of Cell String*

additional capacity. This eliminates the standard approaches to constant-current charging.

Constant-potential charging can be used instead. If the times between successive discharges are very long (>100 hours), a simple constant-voltage charge may be adequate to bring all cells up to charge. However, certain precautions will improve the odds of maintaining a balanced system. First, the charge voltage per cell should be increased as the number of cells in the battery increases. This enhances the probability of all cells getting an adequate voltage to charge. Figure 4-41 indicates the float voltages suggested for use with increased numbers of cells in the string. From an ideal charging voltage of about 2.30 for a single cell at room temperature, the voltage goes up to the 2.35 V/cell value recognized as standard for 6 or 12-volt batteries. A 48-volt battery would then use a float voltage of 2.38. Since this may leave some cells in the overcharge range while others are undercharged, a better approach is to cut the distance between voltage control points down.

For example a 48 volt battery might have its voltage controlled every 12 or 24 volts. This can be done either by separate charger circuitry or by a resistive voltage divider network. Balancing of long-string CP-charged batteries at regular intervals using a constant current charge is also highly recommended. The float voltage should also be controlled to always maintain a minimum trickle charge current of approximately 0.001C at standard ambient temperature. The advent of economical microprocessor-controlled charging systems has allowed many equipment designers to develop creative approaches to charging of long strings.

## 4.2.9.2    Series/Parallel Charging

While series strings (batteries) are routinely used to increase voltage, parallel strings to increase capacity are much less common. However, some common uses for sealed-lead batteries such as many telecommunications applications utilize parallel strings to provide the requisite discharge capacity. Charging batteries in parallel requires use of constant-potential charging since constant-current charging may result in uneven distribution of charging current among the batteries. Charging series-parallel batteries requires particular attention. The precautions suggested in Section 4.1.8 regarding interconnection of paralleled strings should be employed.

### 4.2.9.3   Low-Temperature Charging

In general, charging batteries at moderately low temperatures (–5°C) is relatively straightforward. Either constant-voltage or constant-current may be used although the preference is for a properly temperature-compensated CP charger. Charging at lower temperatures (–20°C) is feasible, but highly inefficient. Again, either constant-voltage or constant-current methods may be used. In either case, high voltages (2.6 volts per cell and up) will be required and venting may occur since more gas may be generated than can be recombined. Often, the result of charging attempts at low temperatures is enough heating within the cell to warm it to a level where charge acceptance is improved.

Charging any battery in ultra-low temperatures is a problematic process. The starved-electrolyte battery, because of its design, survives the cold as well or better than other batteries. It accepts charge in many situations where other batteries will be damaged by charging.

### 4.2.9.4   Elevated Temperature Charging

Charging at high temperatures may be done using either a constant-current or constant-voltage charger. Charging rates and voltages should be adjusted to minimize overcharge using the Tafel curves from Appendix B. Note that self-discharge rates are increased because of the temperatures and trickle charge currents may have to be increased proportionately to accommodate this.

## 4.2.10   CHARGING POWER SOURCES

Because of the diversity of applications in which sealed-lead batteries are used, they are charged from a variety of power sources ranging from clean, pure DC to AC with significant transients and/or wide swings in voltage. Because of the ruggedness of the batteries, they can usually tolerate highly variable power inputs, but, quite often, with some impact on the life expectancy. Maximum battery life comes when the power source combines with the charger to present the battery with a clean, uniform, stable DC input. But, in many applications, the device has a design life short enough that great efforts to extend battery life are not justified. In such cases the quality of the power source can be greatly compromised.

### 4.2.10.1   DC Power Sources

If DC power, provided at the right voltage, is already available, a dedicated charging system is not required. However, great care must be taken to ensure that the voltage available is truly useful for charging. DC systems relying on batteries to replace the primary power often operate at a system voltage closer to the voltage expected from the battery on discharge load than to the voltage necessary to charge the battery.

#### *4.2.10.1.1 Charging from Vehicles*

The ability of the starved-electrolyte battery to accept high initial currents on a constant-voltage charge allow it to be successfully charged from most automotive electrical systems. In most cases, a 12-volt starved-electrolyte battery can be charged directly through the cigar lighter receptacle without additional charger circuitry. However, there are certain restrictions to this use:

- the vehicle must be running and the alternator feeding the electrical system. The starved-electrolyte battery can not be charged directly from the car battery.
- the battery should not be connected as the vehicle is being started. If connected, the battery may try to discharge back into the vehicle electrical system.
- the alternator must provide at least 13.8 volts. Higher voltages, up to 15 volts, are desirable to speed the charge.

In addition to automotive systems, smaller internal-combustion engines on equipment such as lawn mowers are now using electrical start systems powered by batteries. Economic considerations with these devices often dictates a very crude alternator charging system for the battery. The starved-electrolyte battery has a demonstrated record of surviving the adverse environment encountered in these applications.

### 4.2.10.2  AC Power Sources

Most sealed-lead batteries are charged from some form of AC power source. The exact source specification may vary drastically in voltage (110, 220, etc.), frequency (60, 50, 400 Hz), and quality of regulation. Basically, the battery is concerned about only two things, the degree of voltage variation and the amount of ripple passed to the battery. Depending on the AC electrical grid, the amount of voltage variation seen may range from negligible to quite substantial. The effects of differing line voltages can have dramatic effects on chargers, especially taper chargers where the line voltage variation may produce unsatisfactory charge voltages.

In addition to the gross voltage variations attributable to the line voltage, most batteries are exposed to some degree of voltage ripple due to imperfect rectification of the AC current. The worst case is obviously a half-wave rectifier where the battery sees the maximum amount of ripple. While the starved-electrolyte battery can tolerate this, it does impact the battery life. At the other extreme, highly filtered sources may essentially reduce the ripple to zero.

The choice of the degree of filtering is usually an economic one, trading filter cost against reduction in battery life.

### 4.2.10.3  Photovoltaic Sources

In many remote applications, the combination of a battery and photovoltaic charging system are becoming increasingly popular as the source of electrical energy, especially for such items as radio repeaters, weather instrumentation, and navigational aids. The starved-electrolyte battery is particularly well-suited to use with photovoltaic panels because of its high tolerance for overcharge currents. In some cases, the direct output of the solar panel can be used to charge the battery; however, maximum performance and life will be obtained with the use of voltage regulation circuitry.

Photovoltaic cells act basically as constant-voltage sources with highly variable currents that depend on the time of day, weather, etc. With a typical silicon cell capable of delivering about 0.45 volts, cells will have to be connected in series to supply a charge voltage appropriate to charge the battery. The cell area then determines the charge current. Sizing of the collector array is a complex series of trade-offs depending on location, weather, use scenario, array cost, and required battery life. A blocking diode should be used between the battery and the cell array to prevent the battery from discharging through the cells when the light intensity is low.

### 4.2.11  SUMMARY

Starved-electrolyte sealed-lead cells and batteries perform best when properly charged. The charger designer is usually walking the line between the potential performance problems associated with undercharge and the threats to battery life from high overcharge currents. Undercharge can result in either the battery's inability to meet discharge requirements or cumulative "cycling down" that ultimately causes battery failure. High overcharge currents may decrease battery life, either through loss of electrolyte by venting or loss of grid material through oxidation. As indicated in this section, experience indicates that more application problems are caused by undercharging than by overcharging. The high-purity grid used in some advanced sealed-lead batteries helps to minimize the effects of overcharge on grid corrosion.

Starved-electrolyte sealed-lead batteries may be charged by a variety of methods including constant-potential, constant-current, and taper chargers. Constant-potential chargers restore a large fraction of the discharged capacity quickly, but then require longer charging times to fully charge the battery. Constant-current charging at a single current level is best suited for cyclic applications where the battery is not left on charge for long periods. Two-step constant-current chargers are appropriate for most applications. They can use a high-rate current to quickly restore charge and then switch to a lower-rate current in overcharge. Taper chargers, although inexpensive, are, because of their high overcharge currents, suited only to applications not requiring long battery life.

# 4.3   Storage

Virtually all battery applications involve periods of storage where the battery is left unattended and off charge. These may come in the distribution chain prior to first use or they may be part of the duty cycle of the battery. Various forms of batteries have different limits on the amount of storage they will tolerate and require significantly different approaches to the storage process. Careful definition of a storage procedure that matches the needs of the application with the behavior of the battery will do much to ensure the ultimate success of a battery application.

Storage concerns typically fall into two classes. For some applications, the only major interest is in the length of time the battery can survive unattended without damage. For other uses, the battery must not only survive the storage period, but also retain some capacity after storage. Batteries differ significantly in how they perform in these two regards.

Sealed-lead batteries, because of their low self-discharge rates, can, once charged, be stored for long periods of time while retaining useful capacity. In this regard, they are better than either nickel-cadmium or other lead batteries. This may be important for applications where immediate use after storage is an advantage. However, sealed-lead batteries do not tolerate storage in the discharged state like nickel-cadmium batteries. Therefore special attention to length of storage and preparation for storage is advisable. Suggestions for storage of sealed-lead batteries are provided in the following sections.

## 4.3.1   SELF-DISCHARGE

Any charged battery, either primary or rechargeable, tends to lose charge over time. This capacity loss is referred to as *self-discharge*. Other terms used concerning self-discharge are stand loss or retained capacity which is the capacity remaining after self-discharge. In primary batteries the self-discharge rate is slow, often measured in years. In secondary batteries, the rate of capacity loss is normally higher than in primaries, but the loss is usually less important because the batteries can be charged before use.

Self-discharge in lead batteries occurs for two reasons: extraneous reactions occurring within the cell and the inherent instability of Pb and $PbO_2$ in the presence of highly concentrated electrolyte. These factors result in the conversion of the cell's active materials ($PbO_2$ at the positive plate and sponge lead at the negative) into $PbSO_4$. This effectively discharges the cell. The rate of conversion, like all chemical reactions, is temperature dependent. Thus higher temperatures greatly increase the self-discharge of a lead battery.

## 4.3.2   STATE-OF-CHARGE INDICATION

The ability to directly read a battery's state of charge is an extremely helpful tool in planning for storage. In flooded-lead batteries, the traditional measurement of the state of charge is the specific gravity of the electrolyte. Because of the nature of starved-electrolyte cells and batteries, it is not possible to measure electrolyte gravity. Instead, the open-circuit voltage (OCV) of the cell or battery is the primary indicator of state of charge. As seen in Figure 4-42, the relationship between state of

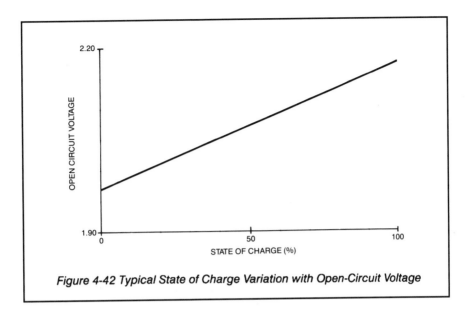

*Figure 4-42 Typical State of Charge Variation with Open-Circuit Voltage*

charge and open-circuit voltage is linear over a small, but usable range of voltages. Experience indicates that, when measured properly and interpreted correctly, the open-circuit voltage is an excellent predictor of remaining battery capacity.

Accurate assessment of battery capacity through OCV requires that the battery have rested from its last charge or discharge. The curve of Figure 4-42 is shown in quantitative form in Appendix B. That curve is accurate within ±10 per cent of the battery capacity if the battery has not been charged or discharged for at least 24 hours.

### 4.3.3    CAPACITY RETENTION DURING STORAGE

For those applications where performance after storage is a major consideration, the important parameter is capacity loss during storage. The decay in OCV with time can be correlated with the curve of Figure 4-42 to produce the capacity loss with time curves of Figure 4-43. Since the discharge reactions are temperature-dependent, capacity loss is greatly accelerated with increases in storage temperature.

Note that these curves were generated for a fully-formed and fully-charged battery. Cells and batteries from some manufacturers are normally shipped at or above an 80 per cent state of charge and become fully formed only after being used. Therefore, the capacity loss in storage prior to first use will be greater than that shown by these curves. Some manufacturers of battery-operated products may want to "boost-charge" their batteries before shipment to take full advantage of the long storage characteristic of the starved-electrolyte battery.

### 4.3.4    CHARGE RETENTION DURING STORAGE

Although capacity loss may be important, the curves of Figure 4-43 substantially understate the maximum amount of time that a sealed-lead battery can survive in storage. This is because there is a "gray area" in the self-discharge of sealed-lead batteries where the battery has no useful capacity, but no damage to battery has occurred.

The key to successful storage of sealed-lead batteries is maintaining a minimum level of charge in the cell or battery. As long as the open-circuit voltage

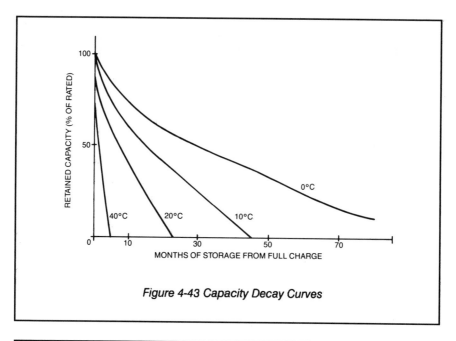

*Figure 4-43 Capacity Decay Curves*

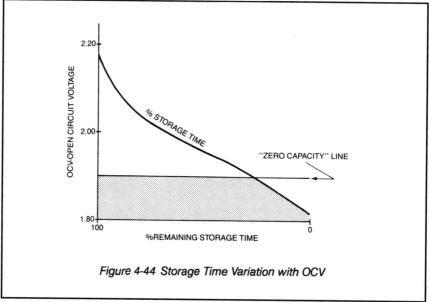

*Figure 4-44 Storage Time Variation with OCV*

remains above this cutoff, the battery does not experience any irreversible changes that affect capacity or life. However sealed-lead batteries should not normally be allowed to self-discharge below 1.8 volts per cell in storage. This value of OCV is significantly lower than the 1.98 volts that represents effectively zero available discharge capacity in the battery.

Figure 4-44 shows storage time remaining versus OCV. As can be seen from comparing Figure 4-42 and 4-44, the battery still has about 60 per cent of its storage life left when it has effectively no useful capacity left. Some manufacturers recommend that, when feasible, batteries in storage receive boost charges at the time they reach the zero-capacity point. This may reduce the need for the capacity-recovery techniques discussed in Section 4.3.5.2.

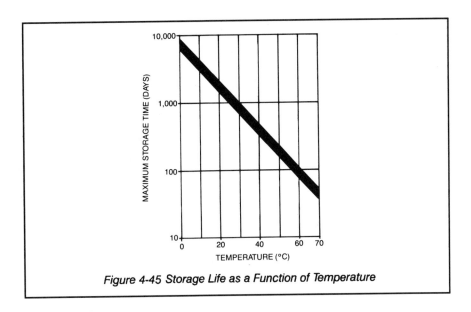

*Figure 4-45 Storage Life as a Function of Temperature*

### 4.3.4.1  Effect of Temperature on Charge Retention

Since the self-discharge reactions are temperature-dependent, the temperature at which a battery is stored will greatly affect the time it can be stored and remain viable. The variation of storage life for a fully charged sealed-lead battery is illustrated in Figure 4-45. At room temperature a fully formed and fully charged battery may last nearly three years before it needs recharging. However, the allowable storage life decreases about 50 per cent for every 10 degrees Celsius that the temperature increases. Of course the reverse effect will benefit batteries stored below room temperature.

### 4.3.5  STORAGE CONDITIONS

In general, storage of sealed-lead batteries is fairly straightforward. A major caveat is that the batteries must be stored in the open-circuit condition. Long periods of storage at even extremely low drain rates may result in permanent damage. When possible, batteries should be stored in a cool and dry environment. They may be stored in any position; there is no need for the batteries to be stored upright. Good inventory practice of using batteries on a first-in, first-out basis will help assure that batteries are used before they have a chance to self-discharge significantly. To maximize the length of time a battery or a battery-containing product may be stored, it should be fully charged at the beginning. Batteries that will be stored for extended periods should undergo regular OCV checks and receive boost charges, either as required or on a regular schedule, such as annually, determined by storage temperature.

### 4.3.5.1  Storage Limits

Since storage temperature greatly affects the speed of the self-discharge reactions, it is the primary determinant of the time a battery may be stored. As indicated earlier, if the goal is to retain some functional capacity, Figure 4-43 indicates the rate at which capacity decays at various temperatures. If, however, the intent is just to avoid permanent damage to the product in storage, Figure 4-45 shows that storage times at all typical temperatures are relatively lengthy.

If cells or batteries are going into long-term storage, an initial charge prior to storage is recommended. A schedule for periodic charges should be developed based on that initial charge date.

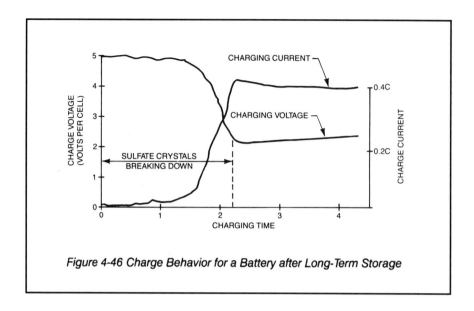

*Figure 4-46 Charge Behavior for a Battery after Long-Term Storage*

### 4.3.5.2    Recovery After Storage

In most normal manufacturing situations, the batteries will go into use well before their OCV drops to 1.98 volts per cell (the zero capacity point).

If cells or batteries have been stored for extended periods which have resulted in their OCV's being significantly below 1.98 volts per cell, although greater than 1.8 volts per cell, the special conditioning described below is recommended prior to use. As cells sit in storage and self-discharge, the active materials convert to lead sulfate ($PbSO_4$) just as they do in other discharges. But, in self-discharge, the lead sulfate forms as larger crystals that have the effect of insulating the particles of the active material, either from each other or from the grid. This result, which is known as sulfation, is normally reversible, but may require a special conditioning procedure prior to use of the battery.

Since sulfation increases the resistance of the cell, the most effective approach to reversing it is to use a low-rate (typically 0.05C) constant-current charge with a supply capable of high charging voltages ($\geq$5 volts per cell). Even though the charger may not be able to force the set-point current through the battery, it normally will be able to force a small current through. As that current works on the lead sulfate crystals, they gradually convert back to active material and the impedance of the cell goes down. Ultimately, the sulfate crystals are broken down and the battery impedance drops to normal levels. Continued charging then returns the cell to its original fully charged state. Figure 4-46 shows typical behavior for a sulfated D-cell. The initial conditioning charge may take as long as 48 hours to break down the sulfation. It is often followed by a standard discharge (0.1C) and a charge to ensure the process is completed.

### 4.3.6    SUMMARY

Starved-electrolyte sealed-lead cells and batteries possess excellent storage characteristics. Their low self-discharge rates provide superior capacity retention, permitting many products to retain an initial operating capability despite extended storage. To get best results on storage, batteries should be charged prior to storage, stored at room temperature or below, and charged prior to becoming deeply discharged.

# 4.4   Battery Life

Starved-electrolyte sealed-lead cells and batteries are designed to provide a trouble-free rechargeable source of electrical power that, in most applications, has a life of many cycles or years of operation.

The useful life of any battery is determined by a combination of the actual discharging and charging history experienced, as well as details of the battery's design and construction. There are a variety of events that can result in a battery failing, either from the effects of age or because of some form of premature failure.

Premature failures may come from some manufacturing defects or from application problems. Careful attention to the manufacturing process and its rigorous outgoing quality control standards produces cells and batteries with an excellent record in their freedom from manufacturing defects. Other sections of this Handbook provide guidance on charging, discharging, and application features that, if followed, will minimize premature failures due to application-related causes. So this discussion will be confined to the effects of the aging process on battery life.

Understanding how and why a battery ages is important to product designers since there are a variety of tradeoffs that a designer may make to prolong battery life. Specifically, battery life may be significantly prolonged by the choice of charger and by the use of low-voltage disconnect circuits, but both may add unacceptable penalties in terms of cost and complexity.

## 4.4.1   APPLICATION DISTINCTIONS

Battery life is, first of all, a function of the type of application: float or cyclic. The critical parameter used to evaluate battery life is even different for the two applications. For float applications where the battery remains ready for service but is rarely used, the parameter of interest is calendar life: "How many years is this battery going to survive?" But, calendar time is normally less important in cyclic applications than the number of cycles. Here the question is: "How many times can I discharge and charge this battery?" The difference between float and cyclic applications was discussed in Section 2. The distinctions between the two types of applications are often more subtle than they first appear. In particular, cyclic versus float is often considered synonymous with portable versus stationary. This can be deceptive. In general, anything that is hard-wired to the electrical grid is extremely likely to be a float application. But other applications are not so clear cut. Household appliances and portable lights are often considered to be cyclic applications, but their typical duty cycles may show long periods on charge coupled with relatively shallow discharges. This implies that they should be considered as float duty for purposes of life analysis, but must be considered as cyclic from the point of view of recharging. The distinctions between float and cyclic applications will be carried through most of the discussions in this section.

### 4.4.1.1   End-of-Life Definition

In general, sealed-lead cells and batteries behave as indicated in Figure 4-47. At the time the cells or batteries are shipped, they are at 80 per cent state of charge or better. In use, either cycling or on float charge, they improve in capacity. After a few cycles,

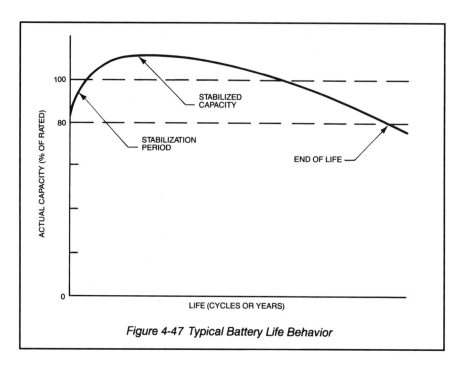

*Figure 4-47 Typical Battery Life Behavior*

they have reached their rated capacity. With continued use, their capacities normally continue to gradually improve until they stabilize at or somewhat above 100 per cent of their rated capacity. The batteries then begin a very gradual decline in capacity. This decline usually continues through the end-of-life point (defined as 80 per cent of rated capacity) and on downward.

The choice of the appropriate end-of-life criterion is a subjective judgment. Some designers choose to define the end of battery life when the capacity is 50 per cent of its rated value. While this does give a longer life projection, it does so with some adverse effects on many battery applications.

For most applications, the parameter that really counts is the end-of-life capacity. Selection of the appropriate battery size must be done based on supplying the load at the end of the battery's life. If the end of battery life is defined as 50 per cent of rated capacity, the battery size required will be double that indicated by use of the rated capacity. If the end of life is specified as 80 per cent of rated capacity, the over-sizing necessary will only be 25 per cent. If the rate of capacity loss per cycle were to be constant between 80 per cent and 50 per cent of rated capacity, the extra battery size would be compensated by an equal increase in battery lifetime. But, since the rate of capacity loss increases between 80 per cent and 50 per cent, the extra capacity does not buy the user an equivalent increase in battery lifetime. Also by designing to a 50 per cent end-of-life criteria, the product is spending a substantial fraction of its life depending on an older battery that is vulnerable to increased failures from mechanical problems.

## 4.4.2   AGING MECHANISMS

To appreciate battery life projections and their supporting rationale, it is useful to know something about the various aging mechanisms that affect sealed-lead batteries as they are used. In general, aging involves the positive plate, either changes in the performance of the active material (changes in plate morphology) or changes that affect the positive grid (grid oxidation).

#### 4.4.2.1   Plate Morphology

As a battery is used, a fundamental change in the structure of the active material on the positive plate occurs. Initially, the $PbO_2$ is in its $\beta$ crystalline form which is highly active electrochemically. As the cell is charged and discharged many times, there is a gradual loss of surface area and some of the $\beta PbO_2$ is converted to an amorphous structure which is electrochemically less active. The progress of this conversion accounts for the bulk of the capacity degradation noticed as the plate ages through cycling.

#### 4.4.2.2   Grid Oxidation

The other major culprit in cell aging is the oxidation of the lead metal in the positive grid to $PbO_2$. This process is primarily a function of the degree of overcharge (amount of current and length of time) experienced by the cell. The oxidation mechanism was discussed in conjunction with overcharging in Section 4.2. The effects on cell life are either mechanical (as discussed below) or electrical. The electrical effect is attributable to the substantially lower electrical conductivity of $PbO_2$ when compared with the metallic lead in the grid. As the conductive cross-section of the grid is reduced, the cell resistance increases, diminishing both the current delivery on discharge and the cell's ability to accept charge on a constant-voltage charge. This gradual loss of capacity must be weighed against the catastrophic failure caused by inadequate or improper charging.

#### 4.4.2.3   Mechanical Deterioration

Mechanical deterioration of the cell is largely confined to the changes in the positive plate due to oxidation of the grid. Since $PbO_2$ occupies a volume about 20 per cent greater than metallic lead, as the positive plate oxidizes, it "grows."

This growth degrades the interface between the plate active material and the current-carrying grid. In conventional lead-acid batteries, the result is shedding of the positive active material where the active material falls off the plate and collects in the bottom of the cell. With the starved-electrolyte cell, the intimate contact between the plate and the separator lessens, but does not eliminate, degradation of the interface with grid oxidation.

Just as the grid grows out, it also grows up vertically. If clearances are too tight, this may result in the positive plate growing into contact with the negative plate terminations resulting in an internal short. Or, if the positive plate is restrained too tightly, it may actually buckle which normally also results in a short.

Lead dioxide ($PbO_2$) has little structural integrity. The ruggedness of many starved-electrolyte batteries comes from the design of the positive plate that uses the separator to compress the $PbO_2$ active material against the grid which supplies the mechanical strength. As the grid is itself converted to $PbO_2$, the plate support becomes increasingly fragile. Thus, as the battery ages, it becomes more vulnerable to failure due to mechanical damage. Typically this manifests itself as loss of electrical continuity for part or all of the plate as the grid fractures due to mechanical shock or vibration.

Mechanical failures of sealed-lead batteries other than those attributable to grid oxidation are exceptionally rare when the batteries are used in normal environments. Fatigue failures of the current-conducting structures may occur in prolonged and/or intense vibration exposures. Contact the battery manufacturer for additional information on applications in nonstandard environments.

### 4.4.3   FLOAT LIFE

The life of the starved-electrolyte battery when used in a float application is one of its major strengths. Because of its ability to recombine the gases evolved during charge and overcharge, it will survive unattended far longer than other lead-acid battery types.

### 4.4.3.1    Life Definition

In float applications, the primary concern regarding life is calendar time, specifically the amount of time spent on overcharge. Most standby power systems are rarely exercised to the point of one deep discharge as often as once a month. Over a ten-year life, this amounts to 120 discharges. As we will see in the cycle-life section, this is well within most batteries' normal cycle life. As a result, concerns over plate morphological changes are usually subordinated to minimizing the rate of grid oxidation while maintaining adequate charging. The different parameters that affect grid oxidation and their effect on float life are discussed below.

### 4.4.3.2    Factors Affecting Float Life

Although there are a variety of factors that may affect the life of the battery in a float application, they all focus on minimizing the overcharge current which in turn minimizes the grid oxidation.

#### 4.4.3.2.1  Charge Current/Voltage

Given that the key to long battery life is minimizing the overcharge current, how is this best accomplished? A reliable approach to charging a battery is the use of a two-step constant-current charger. This ensures a full charge to the battery before switching to a trickle-charge rate that barely replaces the self-discharge losses. This restores the battery to full charge relatively quickly and reduces the problems of cell imbalance. The only concern with this type of charger is that failure of the switching mechanism could leave the battery exposed to the full charge current during extended overcharge. For this reason, use of some form of thermal fuse to remove the charge current if the battery becomes overheated is strongly recommended.

Although the two-step constant-current charger may be the ideal from many charging standpoints, it is less common than constant-voltage chargers in float applications. There are two principal reasons for this: 1) constant-voltage chargers are simple, inexpensive, and reliable, and 2) many float applications involve series-parallel battery strings where constant-current charging is not possible. With constant-voltage chargers, the key determinant of battery life is the charge voltage. To maximize battery life, the charge voltage should be set to produce an overcharge current just greater than the battery self-discharge rate. The Tafel curves, Figure 4-30 and Appendix B, suggest that, at room temperature, a charge voltage of approximately 2.30 volts per cell will provide the minimum amount of charge current. In practice, slightly higher voltages, about 2.35 volts per cell, are recommended to compensate for intercell variations in charge acceptance within battery strings.

It is possible to see the sensitivity of life to charge voltage by examining Figure 4-48. Changing cell voltages from 2.3 to 2.4 volts per cell (a change of about 4 per cent) significantly reduces the projected life at room temperature. The method used to project the data shown in Figure 4-48 is discussed in Section 4.6.

#### 4.4.3.2.2  Battery Temperature

Since the grid oxidation reaction is temperature dependent, increasing the temperature will increase the deterioration process. This occurs no matter what charge current is flowing; however, since the increase in temperature also has the effect of increasing the overcharge current at a given voltage, the result is an intensified effect. The combined effect is seen in Figure 4-48 to result in a greater than 50 per cent decrease in life for every 10°C increase in temperature. This decrease in life may be reduced substantially if the charger is properly compensated.

Obviously the increased temperature's detrimental effect on life is a strong

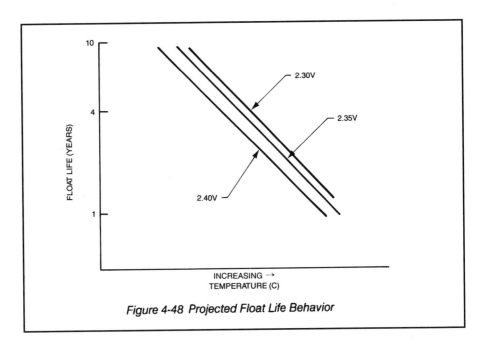

*Figure 4-48 Projected Float Life Behavior*

argument for isolating the battery well away from heat-generating electronic components whenever possible.

### 4.4.3.2.3 Discharge Parameters

Depth of discharge and the time between discharges are not typically major concerns in float duty. Especially for grid-connected applications, it would be extremely rare for a battery to experience a deep discharge (80 to 100 per cent depth of discharge) as regularly as once a month. This type of duty is not likely to impact the life of the battery. The one exception to this would occur if the float voltage were set so low that the battery could not get fully charged and balanced between discharges. Even this is unlikely to be a major concern in grid-connected systems. However, voltage disconnects are still recommended.

## 4.4.4 CYCLE LIFE

Life in a cyclic application—one where the time between discharges is of the same order as the charge time—is a significantly different situation than in a float application. The definition of life changes, the failure mechanisms are altered, and the important variables in determining the battery's life are different. Even though the conditions may be different, starved-electrolyte sealed-lead cells and batteries continue to provide superior performance.

### 4.4.4.1 Life Definition

In a cyclic application, the important variable is not usually calendar time, but number of cycles at a given depth of discharge. While the concerns about overcharge and resulting grid corrosion may still be pertinent, there is an additional factor—changes in the positive plate structure—that must also be considered.

### 4.4.4.2 Factors Affecting Life

The variables affecting life in a cyclic environment are not nearly as simple as those occurring under float conditions. With a float application, the dominant concern was

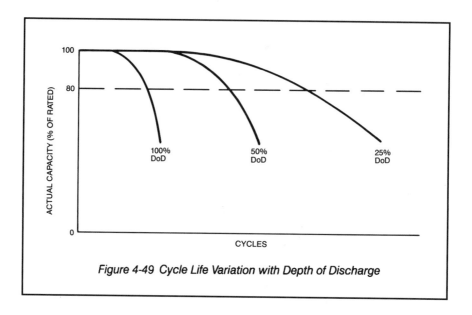

*Figure 4-49 Cycle Life Variation with Depth of Discharge*

overcharge current control. Here that is just one of a variety of concerns. In a cyclic application, performance during discharge and charge affects life as much or more than overcharge.

### 4.4.4.2.1 Depth of Discharge

Far and away, the dominant variable affecting cycle life is the depth of discharge experienced by the battery. The deeper the discharge, the shorter the life. As can be seen in Figure 4-49, the effect of depth of discharge is nonlinear. By decreasing the depth of discharge, the lifetime (number of cycles) can be significantly increased. Battery life for very shallow discharges may approach that seen for float service.

The nonlinearity of cycle life with depth of discharge is the basis for the common suggestion to oversize batteries to extend their life in cyclic applications.

### 4.4.4.2.2 Cycle Time

Cycle time is next to depth of discharge in importance in affecting battery life simply because it determines the latitude that the application designer has in designing charging circuitry. Not just the time between discharges is important, but also whether the cycle is repetitive or sporadic. Even though the battery in a miner's cap lamp and an oscilloscope may be designed to the same duty cycle, the charging scheme to maximize life may be very different. In both cases the batteries may be required to function on an eight-hour discharge and a 16-hour recharge cycle. But, in the miner's cap lamp, that cycle will be repeated essentially every day. Use of the oscilloscope may be far more sporadic, i.e. heavy use for a couple of days and then sitting on charge for a week and then partial use for several days, etc. Different charging schemes may be necessary to gain maximum life in each application.

Figure 4-50 shows that cycle time increases may increase the life of the battery just because the charge time increases. It also illustrates the importance of understanding the application, since selection of the optimum voltage for a 16-hour charge (2.45 volts per cell) may decrease life significantly if the duty cycle actually results in a 28-hour charge. Conversely the voltage choice for a 28-hour charge (2.40 volts per cell) would significantly reduce cycle life if the duty cycle turns out to result in 16-hour charges instead.

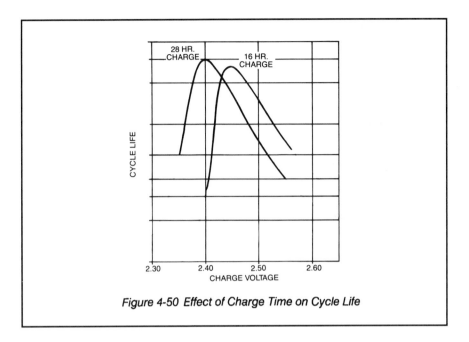

*Figure 4-50 Effect of Charge Time on Cycle Life*

### 4.4.4.2.3 *Charging Parameters*

Unlike float applications where virtually the only concern is overcharge, cyclic applications must also worry about getting enough charge back into the battery. Many designers, faced with incompatibility between quick charge and long life, select the two-step constant-current charger as a good method to handle cyclic applications. This will supply the desired quick charge while accommodating extended periods on charge. Constant-current chargers are also not affected by cycle-down problems.

Constant-voltage (CP) chargers may also be used, albeit with some caution, for cyclic applications. As discussed previously, selecting the voltage for CP chargers must consider the problem of "cycling down" in applications that do not offer opportunities for extended charging to balance cells. This means that charge voltages should be set somewhat higher than in a float application. This reasoning is reinforced by examination of Figure 4-50 where the impact on cycle life of selecting a voltage lower than optimum can be seen to be much greater than that of an equivalently higher than optimum voltage. As a starting point, some manufacturers recommend a cyclic charger voltage of 2.45 volts per cell for typical applications.

### 4.4.4.2.4 *Battery Temperature*

Battery temperature is as detrimental to cycle life as it is for float life. The problem can be reduced significantly by appropriate temperature compensation of the charge voltage. Contact the battery manufacturer for assistance in selecting a temperature compensation approach for a cyclic charger.

## 4.4.5    TRADEOFFS

There are a variety of things that may be done to extend the life of sealed-lead batteries. In developing the design of a battery and charger combination, there may be certain design properties that can be traded off to enhance other desirable product features. Some tradeoffs that may be considered include the following:

### 4.4.5.1   Reliability vs. Life

In many applications it is less important to get absolutely the last drop of life out of a battery than it is to be assured that the battery will perform correctly when needed. This assurance may be gained in a variety of ways. The battery may be retired early, before grid oxidation has proceeded to the point where the battery may fail catastrophically. The battery can be oversized to reduce the depth of discharge. Charge voltages can be set higher than optimum for long life so that proper charge may be ensured thus avoiding failure by cycle down.

### 4.4.5.2   Charger Cost vs. Life-Cycle Savings

It is possible to trade expenditures for chargers against savings in extended battery life. For instance, two-step constant-current chargers can do much to extend the useful life of both cyclic and float applications. Whether the incremental cost of a two-step charger is justified can be determined only from the details of the specific application.

### 4.4.5.3   Product Life vs. Battery Life

Battery and charger designers sometimes get so involved in attempting to develop an optimum solution for long battery life, that the life cycle of the product is overlooked. There is little point in designing a battery to last for eight years if the product is only expected to last three.

## 4.4.6   SUMMARY

By understanding the causes of battery aging, it is possible to design battery and charger combinations that will deliver a long life, whether it is to be used in a float or cyclic application. Many applications are neither truly cyclic nor truly float. In such cases, the conservative approach — to consider the application to be cyclic and ensure that the battery is adequately charged — is normally recommended. For float applications, the approaches to long life are twofold: 1) minimize overcharge currents by controlling charge voltage, and 2) keep the battery away from high temperatures. For cyclic applications, the problem is somewhat more complex because the battery also needs to receive adequate charge during each cycle. A two-step constant-current charger often represents a good approach to charging batteries in cyclic duty. Of course, the above caveats about controlling charge voltage and battery temperature still apply to cyclic applications. A major key to designing battery applications that work effectively is to tailor battery life to the product's needs.

# 4.5 Application Information

Starved-electrolyte sealed-lead batteries and cells have been used in a variety of products ranging from children's toys to scientific experiments on the Space Shuttle. The key to their selection for such a variety of applications is their unmatched combination of economy, ruggedness, and performance. Earlier sections have provided significant detail on the attributes of the product. This section discusses some of application features permitted by these attributes. It also presents a quick summary of typical applications.

## 4.5.1 ECONOMIC CONSIDERATIONS

Batteries are a product whose selection is driven much of the time purely by economics. There are often several batteries that will work in a given application. In most situations, the least expensive battery, when all costs are included, is the best choice. Including all of the costs is the key to making the best selection. For example, initial cost is obviously a major factor. But if the less expensive battery lasts only half as long and requires extensive machine downtime to replace, is it a better choice? The answer to questions like this can be determined only by understanding the economics of these batteries as they are applied to a specific use pattern, operating environment and application requirement. Understanding the economic benefits of any given battery for a specific application requires examining the characteristics and features of the battery and evaluating the degree to which these attributes match the needs of the application. The following sections will consider some of the economic considerations in cyclic duty and standby power applications.

### 4.5.1.1 Economic Features of Cyclic Applications

Because starved-electrolyte sealed-lead batteries can be charged and discharged many times, they are frequently selected for cyclic duty applications.

It is tempting to think of the cost of a battery in terms of its procurement cost only. The procurement cost is often evaluated in terms of cost per watt-hour by dividing the purchase price by the watt-hour rating. Different batteries can be evaluated in terms of dollars or cents per watt-hour. But this simple ratio neglects other important factors including two rather important considerations: actual deliverable energy and battery life. It often will result in selection of a battery that has a higher life-cycle cost.

The actual deliverable energy depends largely upon discharge rate and battery temperature and may differ significantly from the rated conditions. (Sections 4.1 discusses these conditions in greater detail). If the drain rate is quite high or if the battery temperature is quite low in the specific application, the actual deliverable capacity may be dramatically reduced from the rated amp-hours. Starved-electrolyte batteries deliver a high percentage of their rated capacity over a wide range of conditions. The derating needed by starved-electrolyte batteries in low-temperature use is less than other batteries, providing a significant economic advantage for these applications. Hence, the cost per watt-hour for the starved-electrolyte battery (or any other battery) should be evaluated on the basis of actual deliverable energy in the specific application rather than simply using the battery's nominal rating.

The operating life in terms of number of discharge/charge cycles is also an important element to consider. Rechargeable batteries of the same rated capacity when cycled on the same load may provide widely different numbers of acceptable discharge/charge cycles. This makes the cost per watt-hour less relevant. Battery cost for cyclic applications is more properly evaluated as the cost per actual cycle.

To be exact, the costs used for comparison should include much more than the initial costs of the batteries. They should also factor in the costs of the charging system needed by each battery and the costs associated with battery failure including the cost of replacing the battery and any cost to the system (product lost, recreating records, mission failure, etc.) when the battery fails. Since many of these costs occur later in time, standard economic approaches such as discounted cash flow may be used to make all costs comparable.

To summarize, for cyclic duty applications, it is important to compare the costs of different batteries in the actual application. In addition to the initial cost, this comparison should consider the actual deliverable capacities of the batteries at the discharge rate and battery temperature of the specific application; use the projected life of the battery under these conditions to develop a cost per cycle; and ensure that the cost per cycle includes all relevant costs. In many cyclic applications, especially those where reliability is important, starved-electrolyte sealed-lead batteries are an economical choice even though they may not have the lowest purchase price.

### 4.5.1.2 Standby Power Applications

In standby power applications the battery is called upon to discharge only during abnormal situations. When a power outage does occur, the battery must dependably respond to provide the needed power during the emergency. Hence, the reliability of the battery is a very important consideration. As in the discussion for cyclic duty applications, a battery can be evaluated in terms of dollars and cents per watt-hour. Again, it is important to understand and evaluate the battery upon its characteristics in the specific application. The actual deliverable energy under application conditions, not the rated performance, is the critical consideration. It is also important that all evaluations consider the battery performance at the end of life as well as when new. As batteries age, their performance degrades even though the demands on them do not. Therefore all batteries have to be oversized based on initial capacity to ensure that they meet the end-of-life performance requirements. As the degree of performance degradation with age varies widely with batteries, the amount of oversizing required and, thus, the initial battery cost also may vary greatly.

Most standby applications do not see many discharges, so calendar time rather than number of cycles is the important measure of battery life. Operating life in terms of years of service in a float condition impacts the economics of the battery selection. Standby battery costs are normally evaluated in terms of dollars and cents per year of service rather than per cycle as was the case with cyclic applications.

In addition to the initial cost of the batteries, there are a range of other costs that should be factored into the cost comparison as well. The cost of the charger, if required, should be included. The costs of battery failure need to be carefully considered. There are the obvious expenses of the service call and cost of the new battery to replace a failed or expended battery. In addition, there is the cost impact on the overall system of having the battery fail to perform when needed. This is especially important in standby applications where the battery is often backing up a critical piece of equipment. Evaluating the cost of battery failure may be tricky, but it often overshadows other cost components. Again, many of these costs occur at different times so they should be adjusted to reflect their timing using discounted cash flow or the equivalent.

Because the costs of failure are often much greater in a standby application, battery reliability is even more important than in many cyclic applications. Because of their better reliability performance when compared to traditional lead-acid batteries, starved-electrolyte cells and batteries have consistently proven themselves to be the low-cost performer in many situations.

## 4.5.2   CHOICE OF SINGLE-CELL VERSUS MONOBLOC BATTERIES

There are two approaches to starved-electrolyte sealed-lead battery construction: combinations of 2-volt single cells or integral batteries (or monoblocs) which package the cells as a unit in a common container and with intercell connections made internally. Either form is available in a variety of capacities. In many applications, the designer has the choice of using either a single-cell or a monobloc battery. Both batteries may have essentially identical electrical performance and, in fact, may use many of the same internal components. But the differences in packaging of the two product lines suggest different uses. Single-cell batteries are often recommended for applications sensitive to packaging configuration or applications requiring a nonstandard voltage. Most battery sizes are readily available in 6 and 12-volt versions. Other voltages, such as 4, 8, and 24, are sometimes supplied for specialty applications. Integral batteries typically offer better space utilization than an equivalent single-cell battery.

## 4.5.3   PHYSICAL CONSIDERATIONS

In the process of developing the design for a product using a battery, consideration must be given to where and how the battery is mounted, the packaging required to meet the application, what provisions for charging the battery should be made and other physical considerations such as temperature or vibration which might affect battery life or performance. This paragraph discusses the physical considerations of applying sealed-lead batteries to the end product.

### 4.5.3.1   Battery Packaging

The designer has a choice of a variety of ways of packaging the single-cell battery to meet the requirements of the application. Some of the considerations in determining the proper packaging are:

- fit,
- cost,
- ruggedness,
- safety, and
- appearance.

The following discussion provides only a brief introduction to battery packaging. Additional information on packaging options is also available from battery manufacturers.

#### 4.5.3.1.1 *Battery Form Factor and Configurations*

The form factor of a battery depends upon the size and shape of the individual modules and the number of modules in the battery. Batteries comprised of single cells can be assembled into the greatest variety of different configurations. However, even monobloc batteries may be assembled into side by side, end to end, and top to top configurations.

### 4.5.3.1.2 Cases

Most monobloc batteries and many single-cell applications do not use any form of case. With single-cell applications, the design must eliminate the possibility of short-circuiting between the terminals. Battery packs for multi-cell batteries can be furnished with a number of different casing materials and configurations. The case material may be a simple heat-shrinkable plastic sleeve, a rigid plastic tube, a vacuum-formed plastic case or an injection-molded plastic case. In some applications, the battery case may be an integral part of the device.

### 4.5.3.1.3 Interconnections and Terminations

A wide variety of battery interconnections and terminations is available.

Intercell connections on single-cell batteries are normally made with welded metal strips. For special assemblies, the intercell connections may be soldered wire of the appropriate size. In high-vibration applications, braided-strap intercell connectors are sometimes used.

Battery terminations may range from the bare spade terminals through soldered wire leads to polarized connectors. The major concerns here are reliability, economy, assurance of proper polarity, and elimination of short circuits.

## 4.5.3.2    Mounting the Battery

When determining where and how to mount a sealed-lead battery, the application designer can take advantage of its battery's ruggedness and sealed, no-maintenance construction. Mounting constraints, such as keeping the battery upright, away from valuable equipment, and accessible for routine maintenance or replacement, are largely eliminated.

The mounting flexibility of the sealed-lead battery is illustrated by two examples:

- Both aircraft and computer manufacturers integrate starved-electrolyte sealed-lead batteries with other electronic equipment.
- Starved-electrolyte sealed-lead batteries often find application in electric-start walk-behind lawn mowers. This environment combines much of the worst of heat, vibration, and nonstandard mounting positions.

The following sections will describe how best to utilize the flexibility of the sealed-lead battery.

### 4.5.3.2.1 Where to Mount the Battery

The battery may normally be mounted where it is most useful in the product. This involves several considerations. If the product is portable and the battery is relatively heavy, it may be desirable to mount the battery so its weight is evenly distributed for easy handling. In applications where the battery is required to deliver high bursts of power to the load, locating the battery as close to the load as possible will minimize line losses.

Despite the best efforts of the battery manufacturer and the application designer, batteries are often subjected to life-threatening abuse by the end-user. It may make sense to have the battery readily accessible for replacement.

If possible, the battery should be mounted as far away as possible from heat generating components. As pointed out earlier, the higher the average battery temperature, the shorter the battery life. The designer should make every attempt to mount the battery in the coolest spot and/or provide ventilation of some sort whenever temperatures above ambient may be encountered.

#### 4.5.3.2.2 *How to Mount the Battery*

After having located the battery, the designer must then select a method of attaching it securely. Several considerations, such as shock, vibration, temperature, maintainability, access, type of enclosure, weight, etc. affect the choice of attachment method. A relatively immobile battery may simply be located on its own base within the application without special mounting. Other batteries, due to location and external forces, may require special bolts through the case into the mount. Some may be located in a separate pocket or brackets supplied on the product as is the case in many power tools. In any event, the mounting must be compatible with the battery configuration, weight and external forces. In all situations the mounting method must be such that it will not damage the battery or electrical connection.

#### 4.5.3.2.3 *Safety in Mounting the Battery*

Safety in using sealed-lead batteries is discussed in detail in Section 4.7. The information there should be carefully reviewed prior to designing any battery application. Some specific safety considerations in mounting sealed-lead batteries are discussed below. Sealed cells and batteries will vent when the gas pressure within rises due to normal overcharge, accidental misuse, abuse or possible charger failure. Even though battery venting is relatively minimal, all enclosures in which sealed-lead batteries are mounted must be designed so that vented gases will not be retained. If the gases which may be vented from the battery are confined within an external container, there is a possibility of an explosion. Simple venting holes in the outer housing eliminate this concern. In addition, the battery must not be mounted in such a way that it might operate submerged in water. The battery potential is sufficient to electrolyze dirty or salty water thereby forming hydrogen gas at the external positive terminal.

The short circuit current of these batteries is very high which can cause severe burns. The battery should be mounted so that an accidental short circuit cannot occur.

Finally, the battery should be mounted to minimize the possibility of physical damage.

#### 4.5.3.2.4 *Heat Transfer in Mounting the Battery*

Batteries on charge or overcharge generate heat. This is normally not a problem as the heat is rapidly dissipated. However, mounting a battery to minimize cell temperature build-up may be a concern in certain severe applications, particularly when the cells are insulated or enclosed. For battery applications that require operation under one of the following conditions, heat transfer in mounting should receive special attention.

- Extended periods of overcharge at 0.1C or higher
- Fast charge applications
- High discharge current
- Confined surroundings
- High ambient temperatures

A simple test can be conducted by burying a thermocouple in the battery pack and measuring the temperature rise on both charge and discharge. As discussed in Section 4.4, high temperatures will eventually lead to battery failure. Hence, if the temperature rise in the application can be reduced by ventilation or "heat-sinking", the overall battery life will be extended. Many packaging engineers use a large heavy-duty metal strap to clamp the battery to the equipment; the strap also serves to conduct heat away from the battery to the equipment chassis.

## 4.5.4 OPERATING ENVIRONMENT CONSIDERATIONS

Once the battery has been mounted in the equipment, it may be exposed to an environment that can have a drastic effect on the battery's life and performance. The major environmental factors that need to concern designers using batteries are discussed below.

### 4.5.4.1 Operating Temperatures

Batteries operate best in moderate temperature conditions. Extreme operating temperatures are generally detrimental to the sealed-lead battery system, causing either degradation of performance (cold temperatures) or reduction of operating life (high temperatures). While, in certain applications, sufficient performance may be obtained outside these limits, a maximum operating range of –40°C to +60°C is typically recommended.

Operation of these battery systems at high temperature accelerates the nonreversible degradation of performance. As discussed previously, proper temperature compensation of chargers will minimize this degradation. High temperatures also increase self-discharge rates thereby decreasing the time the battery can be stored before a charge is necessary.

Operation toward the low end of the recommended temperature range results in a predictable loss of available capacity, but there is no danger of physical damage to charged sealed-lead batteries from low temperatures. Thus fully charged sealed-lead batteries may be stored without problem at temperatures down to –65°C.

### 4.5.4.2 Relative Humidity

Sealed-lead batteries may be operated in all humidity ranges normally suitable for electrical and electronic components. When operated continuously at 95 per cent relative humidity or higher, the terminals, interconnecting straps and steel cases (as applicable) may show modest cosmetic rusting. The sealed-lead cells and batteries have successfully passed the humidity and salt-spray tests of MlL-STD-810.

### 4.5.4.3 Vacuum or Pressure

Being sealed, starved-electrolyte batteries may be used in either a vacuum or positive-pressure environment. The resealable safety vent operates on differential pressure so that it adjusts to whatever external pressure exists.

### 4.5.4.4 Corrosive Atmosphere

Sealed-lead batteries are constructed so that they will resist corrosion in most environments found in commercial applications. In very severe corrosive environments it is possible for the metal parts to corrode, but in most commercial applications this is not generally a problem. Contact the battery manufacturer for assistance if the ambient environment is expected to be extremely corrosive.

### 4.5.4.5 Shock and Vibration

While sealed-lead batteries offer excellent overall vibration resistance, exposure to prolonged and/or intense vibration may cause premature failure. Most standard product lines are designed to withstand the shock and vibration encountered in routine shipments via common carriers as well as that encountered in most commercial applications. The survival of sealed-lead batteries in walk-behind lawn mowers demonstrates a vibration resistance not found in other lead-acid batteries. There are a few design measures that can be used to reduce the vulnerability of sealed-lead cells and batteries to vibration damage. Two types of vibration-induced failures are seen most often: shearing of the plate tabs and current collectors within the cell or failure of the cell terminals external to the cell.

The internal failure mode is most pronounced when the battery is vibrated in the "vertical" direction. In such situations, the pack containing the plates and separator may start moving relative to the cell container. This movement, if it persists, normally results in fatigue failure of the parts that carry the current from the positive plate to the positive terminal. If the battery can be oriented so it experiences vibration in the "horizontal" rather than "vertical" direction, its vibration resistance is normally increased.

External failure of the terminals is predominately seen with batteries of single cells although batteries of multiple monoblocs may also experience the same difficulty. The problem arises when two cells or monoblocs that are rigidly interconnected begin to move relative to each other under the influence of the vibration. This will normally cause fatigue failure of the terminal or the interconnection. Two methods may be used to reduce or eliminate these failures. One approach is to eliminate cell-to-cell movement within a battery case by either gluing the cells to the case or using an interference fit to hold the cells in the case. A second approach is to provide a flexible interconnect between cells which allows movement between cells without fatigue failure of the electrical linkage. If soldered stranded-wire intercell connections are used in a vibration application, they must be sufficiently long to allow adequate flexibility after the wire has been tinned.

## 4.5.5   TRANSPORTING STARVED-ELECTROLYTE SEALED-LEAD CELLS AND BATTERIES

The unique construction of starved-electrolyte sealed-lead cells and batteries makes them exceptionally easy to transport. With their starved-electrolyte design, they are not subject to the electrolyte spills and corrosion that cause concern with other lead batteries. This advantage has been recognized by various regulatory bodies. Both the U.S. Department of Transportation and the International Air Transport Association have accepted some forms of starved-electrolyte batteries such as those from Gates for shipment as dry batteries. This means that they may be shipped by air in normal packaging without special handling or precautions. Contact the battery manufacturer for supporting documentation.

## 4.5.6   FUNCTIONAL APPLICATIONS FOR SEALED-LEAD BATTERIES

Sealed-lead batteries are being used in many different applications, most of which can be segmented into four functional types:

- Standby battery power
- Engine starting
- Alternate power source
- Portable power

### 4.5.6.1   Standby Battery Power

One excellent application for starved-electrolyte sealed-lead batteries is in systems which require backup or standby power. As electrical and electronic systems continue to replace human and mechanical systems, everyone is increasingly dependent on the continuity of electrical power. The increased importance of electrical service during power outages and momentary interruptions is leading many equipment designers to provide backup power. This backup may be built into the device or it may be an external unit such as an uninterruptible power supply. In addition, some devices are designed to operate only during the period of a power outage, to provide safety for people in situations which might cause alarm or danger as a result of the power interruption. These situations also require a reliable battery, ready to serve upon instantaneous call.

There are three ways a sealed-lead battery can be used in standby power applications:

• Maintain a function during a power failure
• Initiate a function during a power failure
• Provide a "graceful" shutdown during the failure.

The major use of batteries in standby applications is to maintain vital functions despite the loss of power from the grid. An excellent example is the telephone system which uses an array of batteries and engine-generator sets to provide phone service even when the grid power is unavailable. Sealed-lead batteries are used in a variety of telephone installations for backup power. Their size, reliability, ruggedness, and no-maintenance features make them especially well adapted for distributed applications such as switching systems and line concentrators. Sealed-lead batteries are also used in a multitude of other products such as security alarms, computers, and medical equipment that must continue to work during an interruption in grid power. Sealed-lead batteries are especially well-suited for backup that is integral to the product. Because they are clean and maintenance-free, they can be embedded in the unit rather than having to be separated in their own area.

Even though integral backup is becoming more common, many applications choose to use a separate external power system instead. These uninterruptible or switched power systems are increasingly choosing starved-electrolyte sealed-lead batteries because of their reliability and long life.

A good example of the case where the battery initiates a function once the power interruption occurs is battery-powered emergency lighting. Normally the light is off and the charger is keeping the battery fully charged. Once the line power fails, the battery is used to power the light.

In other situations, the basic equipment function is not sufficiently vital to be kept going when the power goes off, but there is a shutdown procedure that should be accomplished to minimize problems with the system. A classic example is seen with many computer systems where records need to be transferred from volatile memory to nonvolatile memory before the system shuts down and the information is lost.

As microprocessors become integral elements of things as simple as toys and household appliances, the importance of limited battery backup increases. Many of these devices use volatile memory to retain instructions or data. They need backup power to retain their memory even when the power is shut off.

Starved-electrolyte sealed-lead batteries are the choice for many designers of equipment requiring standby power because of their very long life, reliability, and simple charging in float applications. The flexibility in mounting these batteries is a significant feature where space is at a premium.

### 4.5.6.2    Engine Starting

Starved-electrolyte sealed-lead batteries have very low internal resistance which permits very large peaks of power for short periods. One application for short, but very high, power delivery is in gasoline engine starting. One application in particular, electric-start walk-behind lawn mowers, is considered by many to be one of the roughest tests of a battery that exists today. Practically everything is wrong. The battery is often mounted where it is exposed to high temperatures from the engine and high vibration from the blade. The charging system is rudimentary. Duty cycles may combine long periods of dis-use with periods of many discharges in a brief amount of time. Starved-electrolyte sealed-lead batteries have been acclaimed as the best batteries available for this application.

### 4.5.6.3 Portable Power

Sealed-lead batteries are often used as the primary power source for portable consumer appliances, tools, lights, and toys.

Duty cycles for many of these items are close to float service. Consumers may use their portable vacuum or rechargeable light once a week and leave it on charge the rest of the time. This type of service is well-suited to constant-voltage float charging using the sealed-lead battery. Strength in charging combined with high current deliveries and acceptable power densities have made sealed-lead cells and batteries economical choices for many portable power applications.

### 4.5.6.4 Alternate Power Sources

In addition to the portable power applications discussed above, sealed-lead batteries are frequently used as an alternate power source to the normal AC line power. These applications are also portable but are normally powered from the AC line through a built-in DC power supply. When the portable feature is desired, the AC line cord is unplugged and the unit is immediately portable. The battery is normally charged from the built-in DC power supply when the unit is plugged in and is usually fully charged, ready to go, when the line cord is unplugged. The sealed lead battery is ideal for these applications because it can be charged continuously using a properly designed constant-voltage charger with little or no reduction in overall life. And because the charge retention of these batteries is excellent, the battery holds its power during long periods of "off" time when the device is not being operated.

A good example of an alternate battery power source application is the portable television set. The battery equipped portable television set is normally operated from AC line power. Whenever the set is plugged into the line, it operates from line power while the battery is charged and maintained at full charge. If someone wishes to operate the TV away from the convenient wall outlet, the battery is ready to do the job. After the use, the battery is automatically recharged once the set is plugged back in the AC power receptacle and is quickly ready for its next outing.

A second example of an alternate battery power source application is in portable instrumentation, such as voltmeters and oscilloscopes. Again, the instrument normally runs from the AC line, but is ready to be removed to a remote location in a fully charged condition.

## 4.5.7 SUMMARY

Starved-electrolyte sealed-lead cells and batteries are the practical answer to battery needs for a diversity of applications. Because of their longer life they can provide low life-cycle costs. They also provide a degree of flexibility in both location and acceptable environments that allows creative product design. Starved-electrolyte sealed batteries are available in both integral (monobloc) and single-cell versions. Standard factory-assembled batteries offer a variety of options in battery shape, size, case enclosure, and electrical terminations. In addition to the standard designs, single-cell batteries can also be assembled into special designs to meet the needs of specific applications. Because sealed-lead cells are clean and rugged, mounting, location, and environmental constraints on their use are minimized. As a result of their performance benefits, starved-electrolyte sealed-lead cells and batteries are used in standby power, engine starting, portable power, and alternate power applications.

# 4.6    **Battery Testing**

The keys to success in designing applications that use batteries are knowing:

- the performance characteristics associated with various batteries,
- understanding how these characteristics will be expressed in the applications's operating environment, and
- designing with these distinctions in mind.

Most designers find that application-related testing is an essential part of the design process. To maximize the return from testing efforts, testing should be carefully planned and should build upon the information available from other sources (such as this Handbook and consultation with the battery manufacturer). This section provides some suggestions on when battery testing may be advisable and on ways to get the most from the testing.

Battery testing is normally performed for two major purposes:

1) battery characterization — studying how the battery is likely to perform under representative conditions and conducting head-to-head comparisons of different candidate batteries. The results are data that the designer can use in selecting and specifying the battery and designing the supporting equipment.
2) product verification testing — ongoing testing that allows verification of battery quality as part of the manufacturing process.

## 4.6.1    CHARACTERIZATION TEST PROCEDURES

Battery characterization tests provide information on battery performance for a specific application. These tests may range from the very simple (a one-shot trial to confirm a specific data point) to the very complex (an elaborate test matrix designed to obtain statistical data that will allow optimization of the battery to the application). Unfortunately, there are few generally accepted test procedures or protocols. Each prospective battery user has to develop test procedures that are justified by the design margins in the product.

It is important in gaining maximum usefulness from these tests that the application parameters be well understood. Such particulars as motor loads, duty cycles, frequency of use, etc. should be carefully defined prior to testing to ensure that the tests are relevant to the actual product as it will be used.

For cyclic applications, the battery characteristics of interest normally consist of the following items, listed in generally descending order of priority:

- discharge performance — can the battery meet the electrical requirements for the specified length of time at the lowest temperatures expected for the application?
- charge acceptance — will the battery recover quickly enough to meet the required use profile? How will that recovery be affected by a continuing series of charge/discharge cycles?
- cycle life — how many charge/discharge cycles can the battery reasonably be expected to survive?
- deep discharge — can the unit gracefully survive being completely discharged before recharge?

- storage life — how long can the product be in the distribution chain and still retain functionality?
- $R_e$ testing — determination of the battery's effective internal resistance allows prediction of the battery's voltage under various current loads.
- mechanical/environmental behavior — will the battery operate properly in the location and orientation proposed for it? Will it survive the mechanical abuse that the unit might experience?

For float applications, the battery characteristics of interest are much the same. Some of the concerns in the cyclic list reflect the fact that these are often portable applications. These issues are not normally as important for float applications and are often replaced with questions about cell balancing and overcharge currents. Testing for float applications normally involves some of the following items, listed in generally descending order of priority:

- discharge performance — can the battery carry the load for the specified length of time at the lowest temperatures specified?
- charge acceptance — will the battery recover quickly enough to meet the required multiple outage scenario, especially at low temperatures? Is the charging adequate to maintain battery balance? What are the overcharge currents, especially in any elevated temperature scenarios?
- float life — how long will the battery survive in this application under the proposed overcharge currents?
- $R_e$ testing — determination of the battery's effective internal resistance allows prediction of the battery's voltage under various current loads.
- mechanical/environmental behavior — will the battery operate properly in the location and orientation proposed for it?

### 4.6.1.1   Preparation for Testing

No matter what characterization testing will be performed, the first step is to develop a properly prepared and conditioned sample of batteries. For various reasons, batteries may arrive for test at different states of charge. Therefore it is important that batteries be prepared for testing by cycling them until their actual capacities are stabilized at a value equal to or greater than the rated capacity.

### 4.6.1.2   Discharge and Charge Acceptance Tests

Often the first set of tests to be performed are those that confirm the successful operation of the battery under nominal conditions, i.e. with the battery new, but stabilized; charged in the optimum manner; and at room temperature. If the battery can not meet the load requirements under favorable conditions, there is no point in continuing. Assuming those tests are favorable, battery operation should be confirmed at the temperature extremes for which the product is expected to operate.

Once the basic ability of the battery to accommodate the load has been determined, then the charge acceptance of the battery can be tested. It is essential that the ability of the proposed charging system to charge the battery within the expected application profile be closely examined. This is where careful testing and feedback to those developing the application requirements will pay substantial benefits to the designer. The result can be substantial cost savings and/or performance improvements through a better understanding of battery behavior.

### 4.6.1.3   Life Tests

As might be expected, the type of life test employed depends on the type of duty the battery will see.

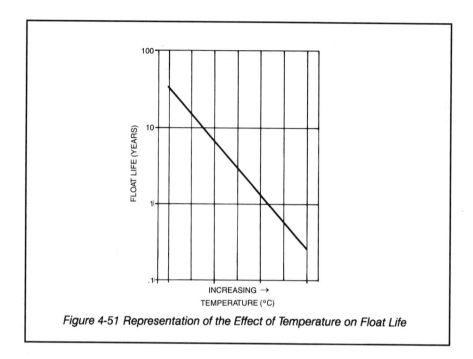

*Figure 4-51 Representation of the Effect of Temperature on Float Life*

### 4.6.1.3.1 Float Life

Battery life for a float application is normally determined by the battery's ability to survive on a constant charge. The major influences on float life are temperature-dependent electrochemical reactions. Therefore, testing at elevated temperatures can, theoretically, be used to project life at normal conditions through the Arrhenius equation. The approach used in the battery industry has been summarized by E. Willihnganz. As indicated in Figure 4-51, testing at 60°C can condense a normal float life of 8 to 10 years to approximately 6 months. Typical tests involve use of well-regulated ovens containing battery samples being charged at the room-temperature float voltage. At monthly intervals, the samples are removed, cooled to room temperature, and given discharge tests.

Use of this form of accelerated testing should be very limited and extrapolation of the resulting data to room temperature applications done very cautiously. The failure mechanisms that limit the life of sealed-lead batteries are not the same at all temperatures. The failure mechanisms at 60 or 70°C are not necessarily the same ones seen at room temperature; in fact, they are often quite different. Accelerated life testing, if used at all, should be limited to providing qualitative information regarding battery performance. Only extended testing under actual conditions will provide definitive information on battery life in a float application.

### 4.6.1.3.2 Cycle Life

Battery life for a cyclic application is normally determined by the number of charge and discharge cycles that the battery can withstand. Unfortunately, testing for cycle life is difficult to perform and the results may be hard to interpret. It can be both expensive and time-consuming. As a result, proposals for cycle testing should be carefully scrutinized prior to committing to the testing.

Normally cycle testing is performed on some form of automated tester that will repeatedly cycle the battery through a nominal charge/discharge cycle. Failure occurs when the battery will not meet the required discharge time. The temptation, of course, is to adopt an abbreviated duty cycle to accelerate the testing. It is extremely difficult

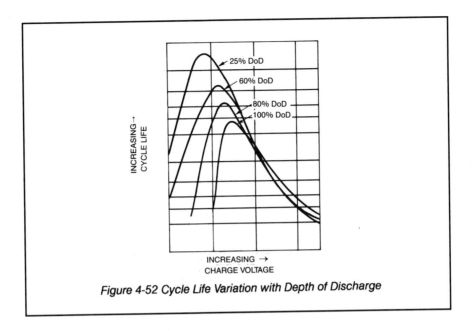

*Figure 4-52 Cycle Life Variation with Depth of Discharge*

to do this and keep the test meaningful. The pacing item is normally the charging time. If the charge time is decreased and the discharge is not decreased, the ratio of charge to discharge is jeopardized and the results may not be meaningful. If the discharge time is also decreased, another problem arises. As shown in Figure 4-52, cycle life is a very nonlinear function of the depth of discharge so that using results at one depth of discharge to predict performance on another discharge is very shaky. Because the test apparatus is expensive, the number of samples that can be cycle tested is usually limited. The result may be the estimation of cycle life by the questionable extrapolation of inadequate quantities of test data. Accelerated cycle testing may be useful to rank or compare the performance of different groups of batteries, but as a tool to predict useful life, it is usually questionable. Only if the proposed cycle can be exactly repeated is cycle testing useful as a prediction of battery cycle life.

### 4.6.1.4 Storage Life

Tests for capacity retention in storage can be conducted using elevated temperature tests similar to those used to estimate float life. Since the pertinent self-discharge reactions in the battery follow the Arrenhius equation, storage time as a function of temperature is of the form shown in Figure 4-53. Thus, by testing batteries for self-discharge at 65°C, three years storage at room temperature can be duplicated in about two months.

### 4.6.1.5 Overdischarge Recovery

For those applications where economics preclude using a disconnect circuit, the battery may be vulnerable to overdischarge by inadvertently being left connected to the load. In this situation, tests that indicate battery performance under deep discharge conditions may be useful. These tests typically involve leaving the battery to discharge through a resistor for an extended period and then requiring the battery to meet a certain discharge standard after one or two charge/discharge cycles. While such tests are instructive, the designer should understand that deep discharge of this magnitude is an abusive condition for many common batteries. He or she should not expect the battery to survive repeated or prolonged deep discharges of this type without adverse effects.

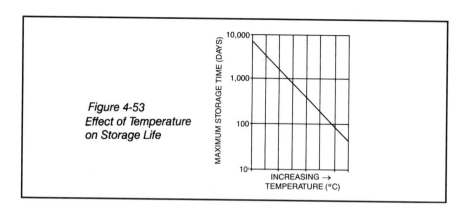

*Figure 4-53*
*Effect of Temperature*
*on Storage Life*

### 4.6.1.6 Effective Internal Resistance, $R_e$, Testing

Numerous measurement methods have been used for determining effective internal resistance of a cell. The value measured on a given cell depends on the method chosen. It is therefore important that, in any comparison of $R_e$ values or in any communication about $R_e$ values, the measurement method be fully defined.

Methods presently in use include AC milliohm meters (1000 Hz), mid-point voltage comparisons at various constant current discharge rates, and the two-current method. The second and third methods yield approximately equal measured values, while the first method results in a very low measured value. The two-current method is endorsed by many manufacturers and battery users because it results in what is felt to be the most useful value and because the measurement can be made very rapidly. The $R_e$ value is used to predict the expected voltage delivery as a function of a constant discharge current. See Section 4.1.4.

The method recommended by many manufacturers is similar to that given in Paragraph 9.4 of ANSI Standard C18.2-1984: *Specifications for Sealed Rechargeable Nickel-Cadmium Cylindrical Bare Cells.* It was selected because it provides a realistic measured value while promoting standardization within the industry.

The test is conducted by discharging the battery at a high rate and then switching to a substantially lower rate. The voltage at the battery terminals and the current is measured immediately before switching and again after the voltage has stabilized after switching. The effective internal resistance is then calculated as:

$$R_e = -\Delta V/\Delta I = (V_L - V_H)/(I_H - I_L)$$

Designers interested in conducting $R_e$ measurements should either refer to the ANSI publication or contact the battery manufacturer for specific recommendations on the testing protocol.

### 4.6.1.7 Mechanical and Environmental Tests

Batteries used in most commercial applications typically have few formal environmental or mechanical requirements imposed. Those requirements that are imposed are often tests to ensure that the batteries can survive normal shipping and customer use. Sealed-lead batteries, both single-cell and monobloc, are rugged enough that these tests are rarely a problem.

### 4.6.2 PRODUCT VERIFICATION TESTS

The characterization testing provides a picture of how the battery should perform. The role of product verification testing is to indicate that the batteries will actually perform as they should.

| Character-istic to be Checked | AQL Suggested for Customer | Reason for Test | Equipment Required | Inspection Limit (At Standard Temperature) | Failure Disposition |
|---|---|---|---|---|---|
| OCV | 1–3 Cell Battery 0.65<br>4–9 Cell Battery 1.00<br>10–19 Cell Battery 2.50<br>20+ Cell Battery requires negotiation | Open or shorted cells<br>Broken welds<br>Polarity | Voltmeter | For single cell batteries:<br>OCV = 2.08<br>For each additional cell, increase OCV by 2.08 volts<br>Example:<br>12 volt battery 6 cells × 2.08 = 12.48 volts | If OCV is lower than limit, boost charge battery at 2.45 volts per cell, let rest for 24 hours, and repeat test.<br>Alternatively run standard capacity test. |
| Capacity | 2.5 | Sufficient run time | Power supply and test fixtures | Per specification requirements at specified temperature | Perform one additional retest to assure proper connections, etc. |
| Battery dimensions | Functional: 1.0 Non-functional: 4.0 | To insure fit | Calipers, micrometers, ht. gage, etc. | Per drawing dimensions | You must insure fit. |
| Weld strength | 2.5 | Weld integrity | Tensile tester | Per specification | Reject |
| Marking, name-plates, date codes | 4.0 | Identification and warranty | Visual inspection | Per specification | Reject |
| Visual workman-ship | 4.0 | Appearance | Visual inspection | Per standard requirements | Reject |

*Table 4-3 Product Verification Reference Guide*

The simplest product verification test is a combination of a visual inspection for obvious problems plus an open circuit voltage measurement. For most lead batteries, the open circuit voltage (OCV) is a good indication of the battery's general health. Many manufacturers have established standards for the OCV as a function of time since charge. Gross deviations from these standards or wide swings in OCV within a lot of batteries may suggest possible battery problems. In such cases a more thorough screening using capacity sampling is indicated.

Capacity sampling is a better, although more expensive, indication than the OCV of the status of a battery lot. Here, a select sample of each battery lot is discharged for residual capacity, charged, discharged again, and given a final charge. The first discharge indicates possible problems in storage while the second discharge can alert the manufacturer to potential battery quality problems.

Table 4-3 indicates recommended product verification tests and acceptance levels.

## 4.6.3    TESTING LOGISTICS

The equipment needed for battery testing is generally typical of that found in any electrical lab. Availability of.power supplies for charge often turns out to be the limiting factor for battery testing. Loads for discharge testing may be either resistors or power supplies used as loads. However, specialized units are available that pro-

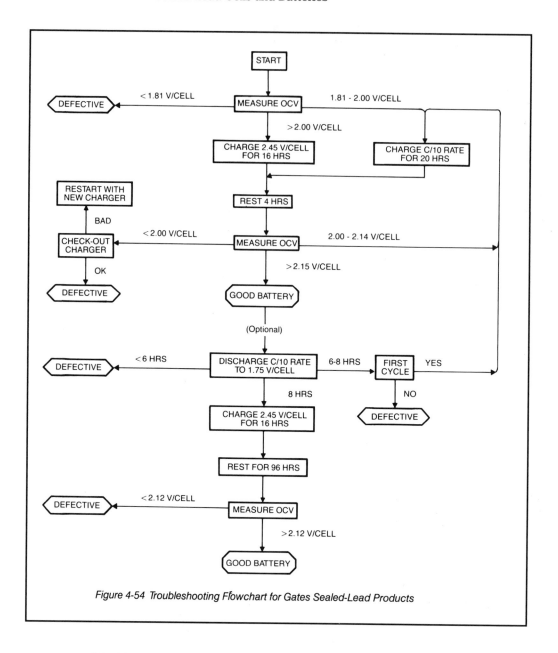

*Figure 4-54 Troubleshooting Flowchart for Gates Sealed-Lead Products*

vide either constant current or constant power discharges adjustable over relatively wide ranges. These programmable electronic loads are fairly economical and do much to improve data quality and increase the speed of data collection.

If cycle testing is to be pursued on a regular basis, an automatic tester is required. These can range from a relatively simple relay-controlled collection of timers and switches to sophisticated switching and monitoring units. For many cyclic applications, either manual cycling or a simple timer/relay-controlled cycle tester may be all that is justified by the resources available, the schedule, or by the sporadic need for repetitive testing. Computerized cycle testing units are mainly appropriate to high-volume product characterization applications.

Scheduling a battery test program can be very challenging and a different experience from most other types of electrical tests. This is due to the long test times often involved. Frequently one test a day is the fastest possible pace. Manpower requirements for this type of testing are often comprised of long periods of inactivity combined with periods of very intense activity.

## 4.6.4 TESTING PRECAUTIONS

In testing batteries, common sense should prevail. Batteries can bite if abused or misused. The safety precautions discussed in Section 4.7 should be followed in any testing.

## 4.6.5 BATTERY TROUBLESHOOTING

There may be occasions where a cell or battery is suspect. In such cases, Gates has found the standard evaluation procedure shown in flow chart form in Figure 4-54 to be useful. This troubleshooting procedure will normally distinguish between batteries suffering from a temporary failure due to the way they have been utilized and batteries that have permanently failed.

The permanent failures can then be evaluated to determine the application parameters that may have resulted in their failure.

## 4.6.6 SUMMARY

Proper application of a battery can substantially increase the effectiveness of the entire system. This often requires knowing specifics of the battery's performance that can only be obtained by testing. Working closely with the battery manufacturer to develop a testing approach can make use of the manufacturer's substantial background and experience to minimize the test effort needed.

If testing is necessary, the key message is to plan ahead. Battery testing is not especially tricky or difficult, but it is time-consuming. The careful and patient designer can get much useful information from battery testing without excessive costs.

It is especially important to carefully examine the interaction between the application profile (charge and discharge) and the actual charging system intended for the product. Although information on battery life is highly desirable, it is also difficult to obtain. Accelerated testing methods are often used to project both float and cyclic life, but interpretation of the results from such tests can be extremely tricky.

# 4.7   **Safety**

---

Starved-electrolyte sealed-lead cells and batteries have compiled an excellent safety record. Over one hundred million cells and batteries have been produced for use in a diversity of both consumer and industrial products. When properly applied, these cells are a valuable, safe source of electrical energy. However, like any other battery type, sealed-lead batteries present possible hazards if mistreated or misapplied. The following sections discuss the various areas of concern and suggest ways of reducing or eliminating possible problems.

There are a number of very important CAUTIONS which should be clearly understood by all who use these batteries, including the OEM, those in the marketing distribution channels and the end user. These CAUTIONS are:

- **These cells contain toxic materials.**
- **Avoid shorting the battery.**
- **Use approved charging methods only.**
- **Do not charge in a gas-tight container.**
- **Properly dispose of used batteries.**

Appropriate caution labels should be placed upon batteries and battery-operated devices.

## 4.7.1   **CELL CONTENTS**

Sealed-lead batteries rely on the interaction between sulfuric acid and various lead compounds to provide electrical energy. These active materials are retained within the plastic cell jar and plastic lid which normally prevent any contact with the cell contents once the batteries leave the factory. However, the cell active materials are toxic and corrosive. Exposure to them should be avoided.

Starved-electrolyte systems operate with the majority of the electrolyte either contained in the plate or adsorbed on the fibers of the separator. There is little free electrolyte to leak if the cell jar is damaged. However, the electrolyte is sulfuric acid in concentrations strong enough to cause serious burns. In case of cell rupture, leaking electrolyte, or other problem that results in SKIN OR EYE EXPOSURE TO THE ELECTROLYTE:

    **IMMEDIATELY FLUSH WITH WATER**

    In cases of contact with skin, consult a physician if burning or redness persist

    If eye exposure occurs, **FLUSH WITH WATER FOR 15 MINUTES** and consult physician.

## 4.7.2   **SHORTING PRECAUTIONS**

Because of its low internal resistance, the sealed-lead cell can provide exceptionally high currents. Normally this is considered a design virtue, but it also means that they can provide dangerous currents when the batteries are inadvertently shorted. Burns to personnel, fires, equipment damage, and battery damage may all occur if cells or batteries are allowed to short. Since sealed-lead batteries are normally maintained in a charged state, even during shipment and storage, care must be taken throughout the battery's life to prevent shorting.

Solutions to this potential problem generally include some combination of battery design and good housekeeping practices.

Several steps can be taken with multi-cell or monobloc batteries to eliminate the possibility of damaging shorts. These precautions include (1) physical design of the battery package to minimize the possibility of simultaneous contact with both terminals; (2) using a current-limiting device such as a fuse or resistor as part of the battery package; (3) terminating the battery with a polarized female connector, and; (4) if necessary, diode protecting the charging leads so that the battery can not discharge through them if they are shorted.

With both cells and batteries, consistent use of insulators to cover battery terminals will prevent problems with shorts.

### 4.7.2.1   Handling Cells and Batteries

Since the terminals of the sealed-lead battery are often upright and exposed, short circuits are best prevented by use of good housekeeping practices as summarized below:

- Whenever possible, keep at least one terminal insulated with a removable cap.
- When working around cells or batteries, always remove rings, watches with metal bands, necklaces or other jewelry which might accidentally complete the circuit.
- Never place a cell or battery in a drawer, trash can, or other receptacle in such a way that it might tip over and short circuit against the walls or that it might short circuit through other metallic contents.
- Do not use uninsulated tools when working with cells or batteries.
- When wiring cells or batteries, beware of the strong affinity of the loose end of the wire for the opposite terminal to the one being wired.
- Never lay any conducting metal part on top of a cell or battery, or place the battery terminals in contact with a metal surface.
- Never place a cell in a pocket containing metal objects such as keys, coins, etc.

### 4.7.2.2   Shipping and Storage

Millions of cells and batteries, properly packed, have traveled millions of miles without incident. However, improperly packed cells or batteries when exposed to the vibration of long-distance truck transportation can short and possibly even ignite the packing materials with unpleasant consequences. The keys to proper shipment are simple:

- Where possible insulate the tabs to prevent contact.
- Use proper packaging materials. Lead cells and batteries are heavy and deserve the protection of adequate strength boxes. Advice on box specifications is usually available from the battery manufacturer.
- Package the product so it cannot move around or tip over. Use styrofoam "popcorn" packaging materials with great care with batteries. Since batteries, because of their weight, tend to "swim" around in the packaging, use of popcorn is advised only for individually protected batteries.
- If stacking cells vertically, remember that cell bottoms are metal and of sufficient diameter to provide an excellent shorting path for the terminals of the cell below. Insulation between layers of cells must resist breaking down under the stress of transportation.
- Boxes of lead cells and batteries are surprisingly heavy for their size. Avoid overstacking boxes of cells or batteries so that the packaging of the lower tier is damaged. Make sure that the freight company does not overload packaging by stacking other cargo over batteries.

If shipping quantities of cells or batteries, consulting in advance with the battery manufacturer on packaging design and shipping procedures can help ensure that the product is received intact.

## 4.7.3    VENTING PRECAUTIONS

Although sealed cells and batteries vent significantly less gas than other forms of lead-acid battery, the gases vented will contain oxygen and/or hydrogen. The vented gases normally diffuse rapidly into the atmosphere; however the mixture of hydrogen and oxygen can be highly explosive if inadequately diluted. Even though venting rates are low during normal charging, **BATTERY ENCLOSURES SHOULD BE DESIGNED WITH ADEQUATE VENTILATION** to prevent an accumulation of explosive gases. **DO NOT CHARGE ANY LEAD-ACID BATTERY, INCLUDING SEALED-LEAD CELLS AND BATTERIES, IN A GAS-TIGHT CONTAINER.**

A second consideration is the potential failure of the charger. If the charger fails, causing higher than recommended charging rates, substantial volumes of hydrogen and oxygen will be vented from the battery which must not be allowed to accumulate.

These cells should never be totally encased in a potting compound, as this prevents the proper operation of the venting mechanism and can lead to dangerous pressure build-up.

## 4.7.4    OVERCHARGE PROTECTION

Certain forms of chargers, especially two-step constant-current, can fail in a manner that causes the battery to see a substantial level of overcharge. This may result in substantial venting from the cell, overheating, and, ultimately, failure of the cells. A thermal fuse mounted so it can sense the batteries being overheated is strongly recommended as part of either the battery or the charger circuitry.

## 4.7.5    DISPOSING OF BATTERIES

Sealed-lead batteries, because they contain lead compounds, have been classified as a hazardous waste that may require special handling depending on applicable federal, state, and local regulations. Organizations needing disposal information should contact their battery manufacturer for advice on the procedures to be followed.

In disposing of sealed-lead batteries, the following precautions should be observed:

- Dispose of these batteries after first discharging them or insulating the terminals to prevent accidental shorting.
- Do not incinerate or expose to fire or high heat, as the cell may burst and spray acid over a large area.
- Disposal of sealed-lead batteries should always conform to applicable regulations.

## 4.7.6    SUMMARY

Starved-electrolyte sealed-lead cells and batteries, when properly applied, are safe and effective power sources. Proper application includes observing the appropriate precautions: The battery should not be confined in a gas-tight environment. The battery should be protected if charger failure could lead to excessive currents. Since sealed-lead cells and batteries may produce high currents if shorted, proper precautions against shorting should be used in packing, shipping, handling, applying, and disposing of the product. The cells contain toxic and corrosive materials; exposure to the cell contents should be avoided. Disposal of sealed-lead cells and batteries should be in full compliance with applicable federal, state, and local regulations.

*Appendix A*

# Sealed Nickel-Cadmium Cell and Battery Product Line Data

This Appendix provides specific data on the Gates sealed nickel-cadmium product line to complement the more general information on sealed nickel-cadmium cells provided within the body of the Handbook. Because Gates is continually changing and improving its product to reflect customer needs, the data presented in this Appendix may not reflect existing production. Consulting with Gates sales engineering and technical marketing personnel early in the design process will provide access to current performance data. Gates personnel can also help meet special needs or resolve application problems.

## CELL PRODUCT LINE OVERVIEW

Gates sealed nickel-cadmium cells are produced in the following general classes:

| Cell Series Designation | Description |
| --- | --- |
| G | VALUMAX™—Cells designed for typical applications. Capable of recharge at rates up to C/3 and continuous use (discharge, charge, and overcharge) in temperatures up to 50°C. |
| GF | VALUMAX Fast Charge—Cells for use in fast-charge applications (with charge termination controls). Capable of recharge at rates up to 1.2C and continuous use (discharge, charge, and overcharge) in temperatures up to 50°C. |
| GX | GEMAX™—Gates maximum power product designed for high capacity and high discharge rate applications. Capable of recharge at rates up to C/3 and continuous use (discharge, charge, and overcharge) in temperatures up to 50°C. |
| GXF | GEMAX Fast Charge—Cells that combine the discharge benefits of the GEMAX product with accelerated charge (using a charge-termination charger) features. |
| GH | Goldtop®—Cells for use in high temperature applications. Capable of recharge at rates up to C/3 and continuous use (discharge, charge, and overcharge) in temperatures up to 70°C. |
| GXH | GEMAX Goldtop—Products possessing the high-performance of the GEMAX family, but capable of use in environments up to 70°C. |
| GHC | Goldtop Plus—Cells that offer extended life at high temperature while providing high-temperature charge acceptance and capacity delivery. |

GM                    MP2 Memory Preservation Goldtop Cells—Cells designed for applications such as computer memory backup where extended high-
temperature life and survival under loaded storage are significant
benefits.

Table A-1 provides a quick summary of the performance features of the various cell
classes.

## ORGANIZATION OF THE APPENDIX

The remainder of the appendix is organized as follows:

### Table A-1 Product Selection Guide
A general review of the different classes of cells that are available from Gates to
meet special operating requirements.

### Figure A-1 Cell Capacity Selector Guide
A nomogram which allows quick preliminary assessment of the cell size that will
best satisfy a specific uniform drain rate and discharge time requirement.

## TYPICAL DISCHARGE PERFORMANCE CURVES

These curves provide voltage profiles for room-temperature discharges at representative drain rates. A family of curves is provided for each class of cells:

Figure A-2    Discharge Performance for VALUMAX™ (G Series) Cells
Figure A-3    Discharge Performance for VALUMAX Fast Charge (GF Series) Cells
Figure A-4    Discharge Performance for ULTRAMAX™ (GL Series) Cells
Figure A-5    Discharge Performance for ULTRAMAX Fast Charge Cells
Figure A-6    Discharge Performance for GEMAX™ (GX Series) Cells
Figure A-7    Discharge Performance for Goldtop® (GH Series) Cells
Figure A-8    Discharge Performance for MP2 Memory Preservation Goldtop (GM)
              Cells

## SCALING THE DISCHARGE CURVE

The process of developing a discharge curve for a specific application was described
in Section 3.1.4. This involves adjusting the rated capacity by application of derating
factors based on discharge conditions to obtain actual capacity. The derating curves
for capacity are provided here:

Figure A-9    Effect of Cell Charge Temperature on Actual Capacity
Figure A-10  Effect of Cell Discharge Temperature on Actual Capacity
Figure A-11  Effect of Discharge Rate on Actual Capacity

Section 3.1.4 also describes the process for estimating the midpoint voltage for the
discharge by derating it based on cell temperature effects on cell resistance and mid-
point voltage. The curves necessary to obtain these derating factors are:

Figure A-12  Effect of Cell Discharge Temperature on Resistance
Figure A-13  Effect of Cell Discharge Temperature on Midpoint Voltage

Once the actual capacity and the midpoint voltage have been calculated, the standard discharge curve for the specific discharge conditions may be generated using the curve shape of :

Figure A-14  Sealed Nickel-Cadmium Cell Universal Discharge Curve

## OTHER CELL PERFORMANCE DATA

In addition to the discharge information discussed above, there are a variety of additional curves that are useful in designing with sealed nickel-cadmium cells. Two particularly important general curves are:

Figure A-15  Cell Voltage vs. Overcharge Current (Tafel Curve)
Figure A-16  Retained Capacity

Useful charge acceptance characteristics are provided by:

Figure A-17  Effect of Temperature on Charge Acceptance
Figure A-18  Effect of Charge Rate on Charge Acceptance
Figure A-19  Effect of Temperature on Capacity
Figure A-20  Time to Return Capacity at Different Temperatures

Cell response characteristics, specifically the behavior of voltage, temperature, and pressure as the cell charges, are useful in designing charging schemes:

Figure A-21  Charge Response Profiles

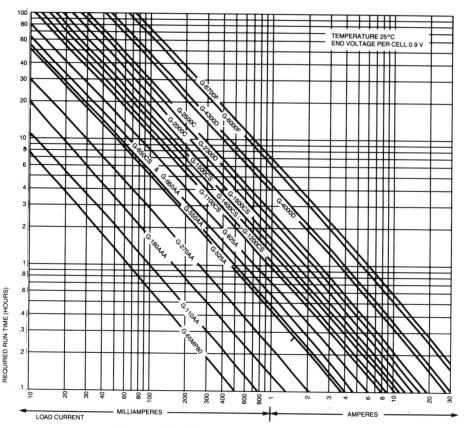

*Figure A-1 Cell Capacity Selector Guide*

## Use of Selector Guide

1. Locate the room-temperature discharge current for your application on horizontal axis.
2. Choose the desired run time on the vertical axis.
3. The intersection of the run time and the discharge current defines the necessary capacity. The required capacity is indicated by the nearest diagonal line that is above and to the right of this intersection. This line is identified by the model number which provides a reference for the cell specifications.
4. Any cell that is above or to the right of the intersection will meet the cell discharge requirement at room temperature. If the application requires a cell with special characteristics such as operation at high temperature, refer to the cell specification.

**Example:** The application requires a discharge current of 300 mA for 5 hours. **Solution:** The run time and discharge current intersect on the capacity line marked "G-1500CS." This model number indicates a $C_s$ diameter cell from Gates standard product line would meet the minimum capacity required for this application. Other cell options are indicated on the chart (anything above and to the right).

Note: This chart uses nominal room-temperature parameters. Careful attention must be paid to various derating factors such as charge temperature, discharge temperature, etc. which may cause the cell to deliver less capacity. If this occurs, a larger cell may be required. The derating parameters indicated in Figures A-9 through A-11 may be useful in assessing this effect. Consult Gates Energy Products for additional assistance in determining the appropriate cell size and design for specific applications.

## Nickel-Cadmium Data A

**Product legend**

- **G** — VALUMAX™ Gates NiCd Cells (G Series)
- **GF** — VALUMAX Fast Charge Cells (GF Series)
- **GL** — ULTRAMAX™ (GL Series)
- **GLF** — ULTRAMAX Fast Charge Cells (GLF Series)
- **GX** — GEMAX™ NiCd Cells (GX Series)
- **GXF** — GEMAX Fast Charge Cells (GXF Series) XP60
- **GTF** — TORQUEMAX™ Fast Charge, High Rate Discharge Cells (GTF Series)
- **GH** — Goldtop® High-Temperature Cells (GH Series)
- **GHC** — Golftop Plus Cells (GHC Series)
- **GM** — MP2 Memory Preservation Cells (GM Series)

| | Product Type | G | GF | GL | GLF | GX | GXF | GTF | GH | GHC | GM |
|---|---|---|---|---|---|---|---|---|---|---|---|
| **Features** | Leakage Protection | | | | | | | | | | ■ |
| | Improved High-Temperature Charge Acceptance | | | | | | | | | ■ | |
| | Improved High-Temperature Life | | | | | | | | ■ | ■ | |
| | Faster Recharge | | ■ | | ■ | | ■ | ■ | | | |
| | Improved Power Delivery | | | | | ■ | ■ | ■ | | | |
| | Longer Run Times | | | ■ | ■ | ■ | ■ | | | | |
| | Value Priced | ■ | | | | | | | | | |
| | Maintenance-Free Operation | ■ | ■ | ■ | ■ | ■ | ■ | ■ | ■ | ■ | ■ |
| | Rugged Construction | ■ | ■ | ■ | ■ | ■ | ■ | ■ | ■ | ■ | ■ |
| **Applications** | Security Alarm and Emergency Lighting | | | | | | | ■ | | ■ | |
| | Telecommunications | | ■ | | | ■ | ■ | | | | |
| | Portable Lighting | ■ | | ■ | ■ | ■ | | | | | |
| | Portable Appliances and Tools | ■ | ■ | | | ■ | ■ | ■ | ■ | ■ | |
| | Consumer Electronics | ■ | | ■ | ■ | ■ | | | | | |
| | Portable Electronic Equipment | ■ | ■ | ■ | ■ | ■ | ■ | ■ | | | |
| | Computer Power | ■ | ■ | ■ | ■ | ■ | ■ | ■ | | | |
| | Computer Memory Backup | | | | | | | | | | ■ |

*Table A-1 Product Selection Guide*

## Nickel-Cadmium Data A

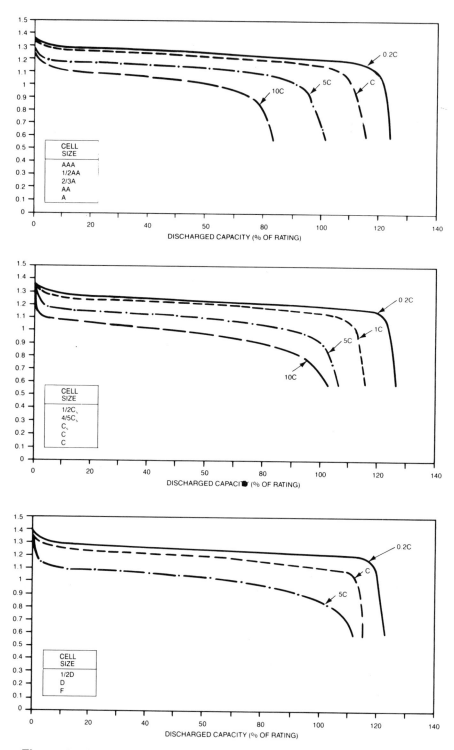

**Figure A-2 Discharge Performance for VALUMAX™ (G Series) Cells**
(Reference: Section 3.1.3 of text)

## Nickel-Cadmium Data A

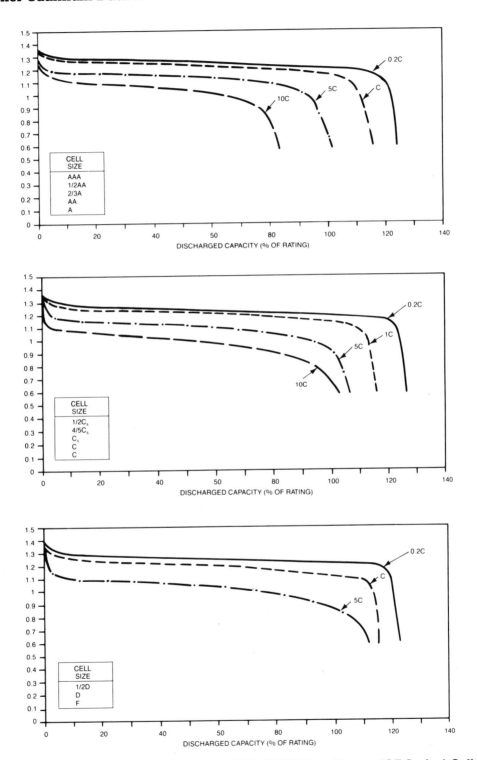

**Figure A-3 Discharge Performance for VALUMAX Fast Charge (GF Series) Cells**
(Reference: Section 3.1.3 of text)

**Nickel-Cadmium Data A**

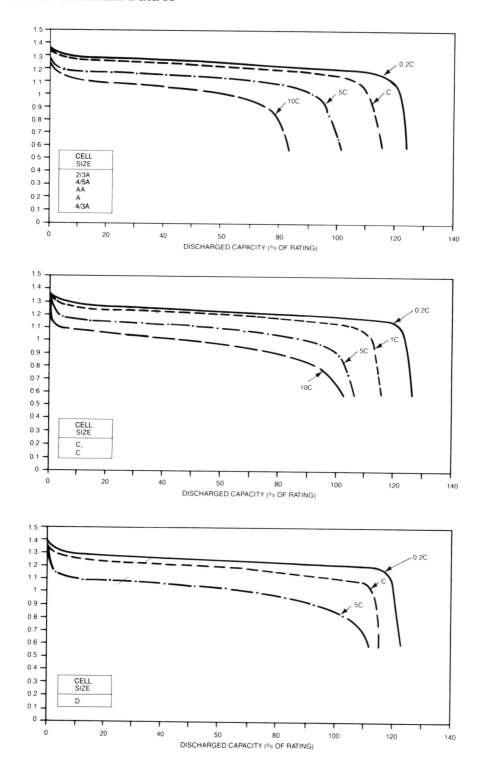

**Figure A-4 Discharge Performance for ULTRAMAX™ (GL Series) Cells**
**(Reference: Section 3.1.3 of text)**

## Nickel-Cadmium Data A

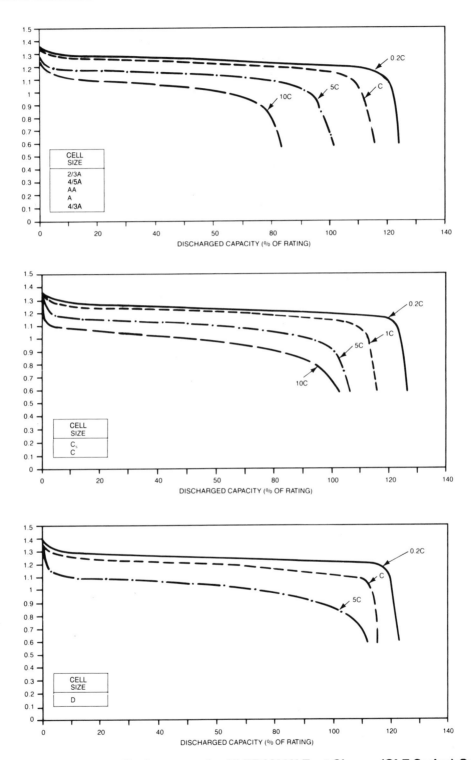

**Figure A-5 Discharge Performance for ULTRAMAX Fast Charge (GLF Series) Cells**
(Reference: Section 3.1.3 of text)

## Nickel-Cadmium Data A

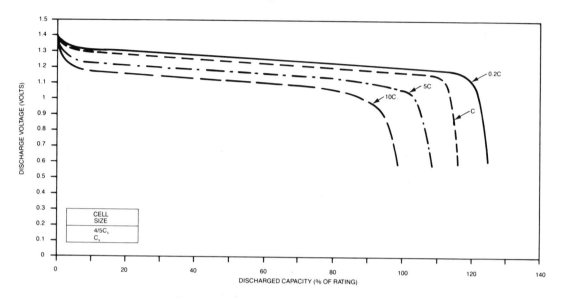

**Figure A-6 Discharge Performance for GEMAX™ (GX Series)
and GEMAX Fast Charge (GXF Series) Cells**
(Reference: Section 3.1.3 of text)

## Nickel-Cadmium Data A

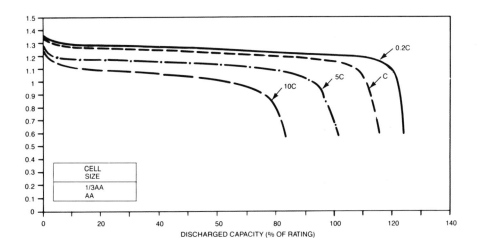

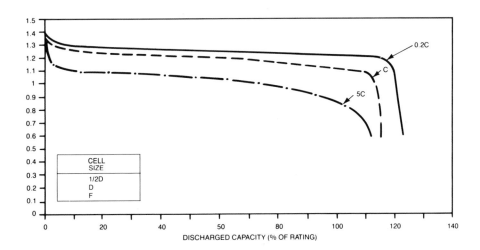

**Figure A-7 Discharge Performance for Goldtop® (GH Series) Cells**
**(Reference: Section 3.1.3 of text)**

## Nickel-Cadmium Data A

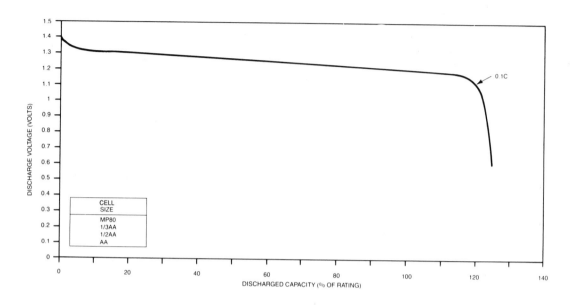

**Figure A-8 Discharge Performance for MP2 Memory Preservation (GM) Cells**
**(Reference: Section 3.1.3 of text)**

## Nickel-Cadmium Data A

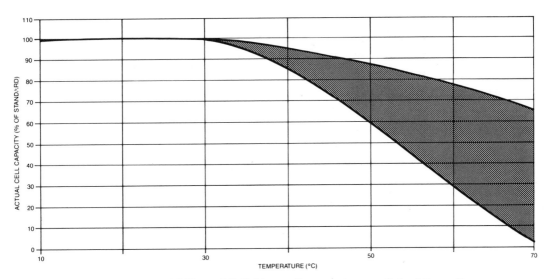

**Figure A-9 Effect of Cell Charge Temperature on Actual Capacity**
(Reference: Section 3.2.2.1 of text)

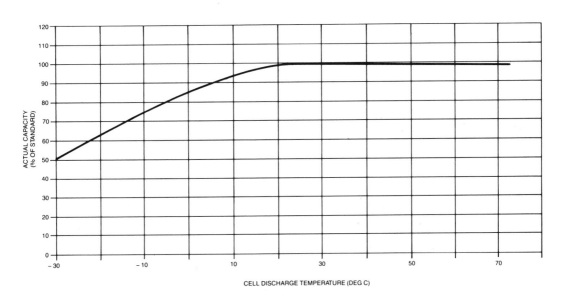

**Figure A-10 Effect of Cell Discharge Temperature on Actual Capacity**
(Reference: Section 3.1.3.3.3 of text)

## Nickel-Cadmium Data A

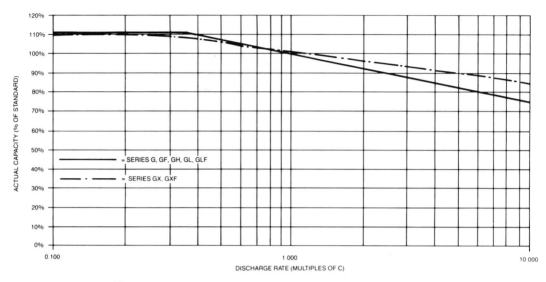

**Figure A-11 Effect of Discharge Rate on Actual Capacity**
(Reference: Section 3.1.3.3.3 of text)

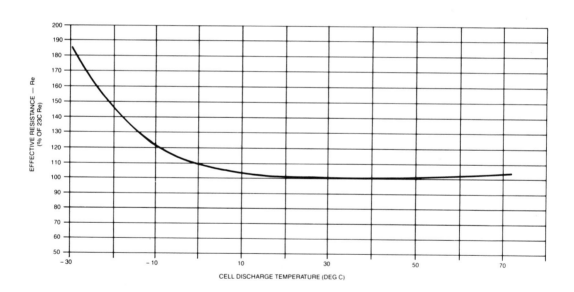

**Figure A-12 Effect of Cell Discharge Temperature on Resistance**
(Reference: Section 3.1.2.3.1 of text)

## Nickel-Cadmium Data A

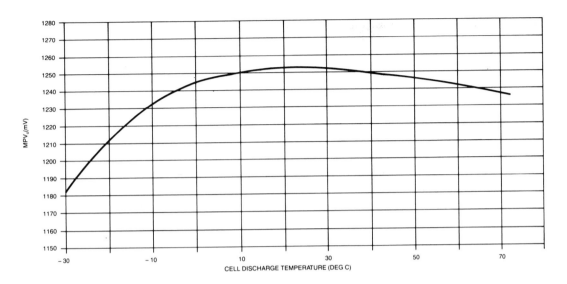

**Figure A-13 Effect of Cell Discharge Temperature on Midpoint Voltage**
(Reference: Section 3.1.2.4.1 of text)

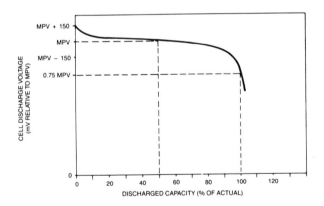

**Figure A-14 Sealed Nickel-Cadmium Cell Universal Discharge Curve**
(Reference: Section 3.1.4 of text)

**Nickel-Cadmium Data A**

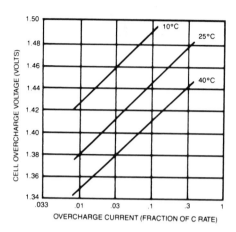

**Figure A-15 Cell Voltage vs. Overcharge Current (Tafel Curve)**
**(Reference: Section 3.2.4.1 of text)**

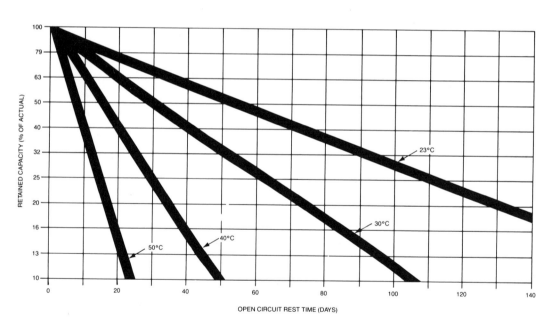

**Figure A-16 Retained Capacity**
**(Reference: Section 3.1.3.4 of text)**

## Nickel-Cadmium Data A

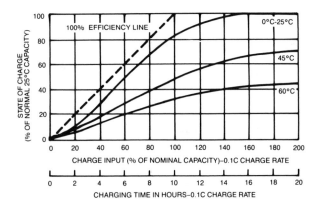

**Figure A-17 Effect of Temperature on Charge Acceptance**
(Reference: Section 3.2.2.1 of text)

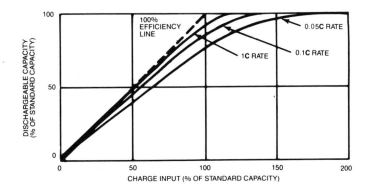

**Figure A-18 Effect of Charge Rate of Charge Acceptance**
(Reference: Section 3.2.2.2 of text)

**Nickel-Cadmium Data A**

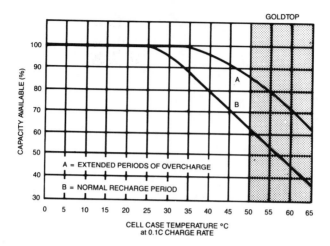

**Figure A-19 Effect of Temperature on Capacity**
(Reference: Section 3.2.2.1 of text)

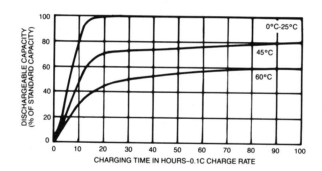

**Figure A-20 Time to Return Capacity at Different Temperatures**
(Reference: Section 3.2.2.1 of text)

## Nickel-Cadmium Data A

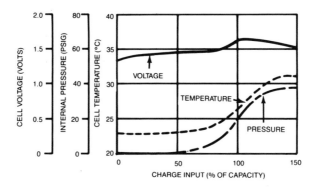

*a. 0.1C Charge Rate*

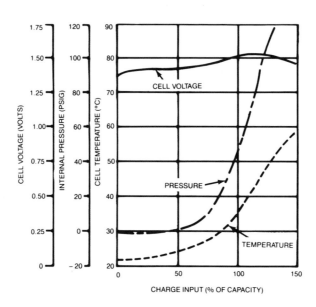

*b. 1C Charge Rate*
**Figure A-21 Charge Response Profiles**
(Reference: Section 3.2.3 of text)

# Sealed-Lead Cell and Battery Product Line Data

This Appendix provides specific data on Gates sealed-lead product line to complement the more general discussion of wound cylindrical sealed-lead cells and batteries within the body of the Handbook. Because Gates is continually changing and improving its products to reflect customer needs, the data presented in this Appendix, although accurate at the time of publication, may not reflect existing production. Consulting with Gates sales engineering and technical marketing personnel early in the design process will provide access to current performance data. Gates personnel can also help meet special needs or resolve application problems.

## CELL PRODUCT LINE OVERVIEW

Gates sealed-lead cells are produced in the two general classes:

| Cell Series Designation | Description |
|---|---|
| Single Cells | Individual 2-volt cylindrical cells which may be designed into a variety of battery configurations. |
| Monoblocs | Modules consisting of two or three cells designed to meet many common application needs. |

## ORGANIZATION OF THE APPENDIX

The remainder of the appendix is organized as follows:

*General*
Figure B-1    Cell Capacity Selector Guide
Table B-1    Product Line Data

*Typical Discharge Performance Curves*
Figure B-2    Discharge Performance for Sealed-Lead Cells and Batteries
Figure B-3    Effect of Temperature on Capacity
Figure B-4    Current-Voltage Relationships
Figure B-5    Instantaneous Maximum Power Curves

*Charging Information*
Table B-2    Charging Options for Sealed-Lead Batteries
Figure B-6    Temperature Compensation for Constant-Voltage Charging
Figure B-7    Typical Voltage Profiles for Constant-Current Charging
Figure B-8    Cell Voltage vs. Overcharge Current (Tafel Curve) Storage Data
Figure B-9    Open-Circuit Voltage versus State of Charge

## Use of Selector Guide

1. Locate the room-temperature discharge current for your application on the vertical axis.
2. Choose the desired run time on the horizontal axis.
3. The intersection of the run time and the discharge current defines the necessary capacity. The required capacity is indicated by the nearest diagonal line that is above and to the right of this intersection. This line is identified by the model number which provides a reference for the cell specifications provided in Table B-1.
4. Any cell that is above or to the right of the intersection will meet the cell discharge requirement at room temperature.

**Example:** The application requires a discharge current of 1 amp for 4 hours.
**Solution:** The run time and discharge current intersect below the capacity line marked "X(5.0 Ah)". This means that Gates sealed-lead X cells and monoblocs would meet the minimum capacity required for this application. Other cell options are indicated on the chart (anything above and to the right) and in Table B-1.

Note: This chart uses nominal room-temperature parameters. Careful attention must be paid to various derating factors such as charge temperature, discharge temperature, etc. which may cause the cell to deliver less capacity. If this occurs, a larger cell may be required. Consult Gates Energy Products for additional assistance in determining the appropriate cell size and design for specific applications.

## Sealed-Lead Data B

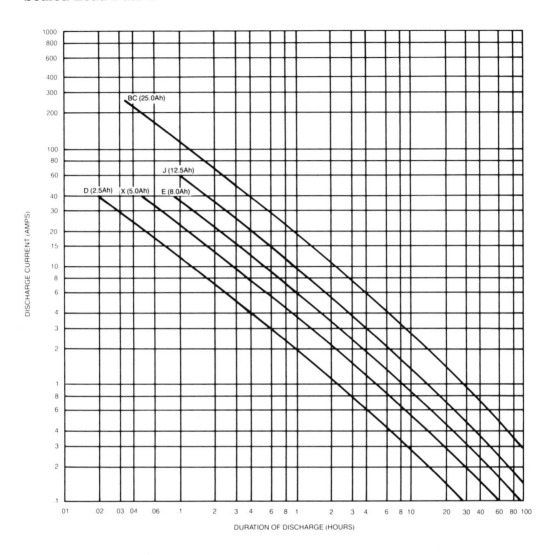

*Figure B-1 Cell Capacity Selector Guide*

## Sealed-Lead Data B

| Model | Capacity | | | Dimensions | | | Weight | Permissible Temperature Ranges | | | | | | Maximum Discharge |
|---|---|---|---|---|---|---|---|---|---|---|---|---|---|---|
| | **C/10** Min | **C/20** Min | **1C** Min | Height mm/in | Diameter Width mm/in | Length mm/in | Typical kg/lb | Storage | | Discharge | | Charge | | |
| | (amp-hr) | | | | | | | Min (°C) | Max | Min (°C) | Max | Min (°C) | Max | (A) |

**SINGLE CELLS** *Proven Gates construction for versatile packaging*

| Model | C/10 | C/20 | 1C | Height | Diameter/Width | Length | Weight | Stor Min | Stor Max | Dis Min | Dis Max | Chg Min | Chg Max | Max Discharge |
|---|---|---|---|---|---|---|---|---|---|---|---|---|---|---|
| D | 2.5 | 2.7 | 1.8 | 67.3/2.65 | 33.9/1.34 | N/A | 0.18/0.40 | -65 | 65 | -65 | 65 | -40 | 65 | 40 |
| X | 5.0 | 5.4 | 3.2 | 80.3/3.16 | 44.1/1.74 | N/A | 0.39/0.81 | -65 | 65 | -65 | 65 | -40 | 65 | 40 |
| J | 12.5 | 13.0 | 9.0 | 135.7/5.34 | 51.7/2.04 | N/A | 0.84/1.85 | -65 | 65 | -65 | 65 | -40 | 65 | 60 |
| BC | 25.0 | 26.0 | 17.5 | 172.3/6.79 | 64.8/2.55 | N/A | 1.58/3.49 | -65 | 65 | -65 | 65 | -40 | 65 | 250 |

**MONOBLOC BATTERIES** *Preassembled batteries in common sizes*

| Model | C/10 | C/20 | 1C | Height | Diameter/Width | Length | Weight | Stor Min | Stor Max | Dis Min | Dis Max | Chg Min | Chg Max | Max Discharge |
|---|---|---|---|---|---|---|---|---|---|---|---|---|---|---|
| D - 4 Volt | 2.5 | 2.7 | 1.8 | 70./2.75 | 45./1.78 | 78./3.09 | 0.36/0.8 | -65 | 65 | -65 | 65 | -40 | 65 | 40 |
| D - 6 Volt | 2.5 | 2.7 | 1.8 | 70./2.75 | 45./1.78 | 113./4.43 | 0.54/1.2 | -65 | 65 | -65 | 65 | -40 | 65 | 40 |
| X - 4 Volt | 5.0 | 5.4 | 3.2 | 77./3.02 | 54./2.11 | 96./3.78 | 0.74/1.62 | -65 | 65 | -65 | 65 | -40 | 65 | 40 |
| X - 6 Volt | 5.0 | 5.4 | 3.2 | 77./3.02 | 54./2.11 | 139./5.47 | 1.11/2.43 | -65 | 65 | -65 | 65 | -40 | 65 | 40 |
| E - 4 Volt | 8.0 | 8.6 | 5.8 | 102./4.00 | 54./2.11 | 96./3.77 | 1.11/2.43 | -65 | 65 | -65 | 65 | -40 | 65 | 40 |
| E - 6 Volt | 8.0 | 8.6 | 5.8 | 102./4.00 | 54./2.11 | 139./5.47 | 1.67/3.65 | -65 | 65 | -65 | 65 | -40 | 65 | 40 |

*Table B-1 Product Line Data*

**Sealed-Lead Data B**

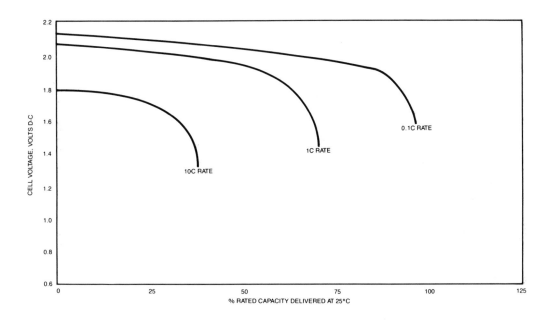

***Figure B-2 Discharge Performance for Sealed-Lead Cells and Batteries
(Reference: Section 4.1.5.2 of text)***

## Sealed-Lead Data B

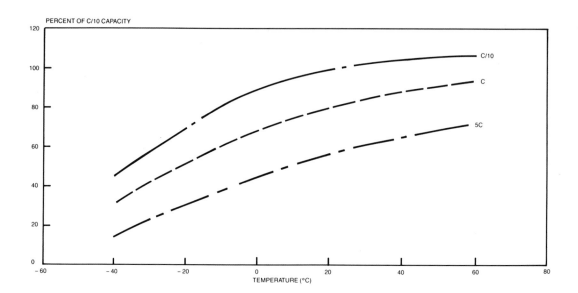

**Figure B-3 Effect of Temperature on Capacity**
*(Reference: Section 4.1.5.3 of text)*

**Sealed-Lead Data B**

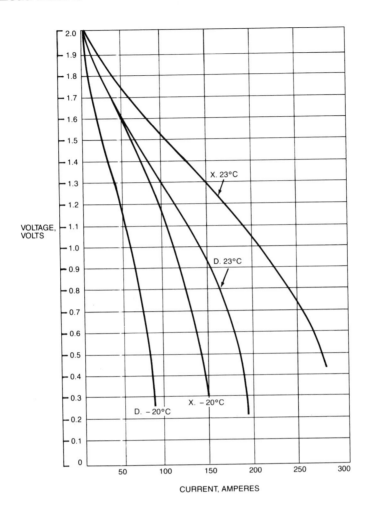

**Figure B-4 Current-Voltage Relationships**
**(Reference: Section 4.1.6 of text)**

## Sealed-Lead Data B

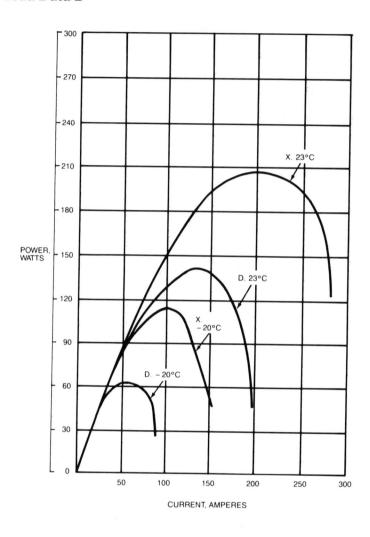

*Figure B-5 Instantaneous Maximum Power Curves*
*(Reference: Section 4.1.6 of text)*

## Sealed-Lead Data B

|  | Standby | Daily Cycle<br>(16 hrs for recharge) | Weekly Cycle<br>(64 hrs for recharge) | Irregular Cycle |
|---|---|---|---|---|
| **Constant Voltage** | 2.30 - 2.35 volts/cell | 2.40 - 2.45 volts/cell | 2.35 - 2.40 volts/cell | 2.35 - 2.40 volts/cell |
| **Constant Current** | - Two-step<br>- Continuous<br>(**C**/500) | - Two-step<br>- Continuous<br>(**C**/14 with charge termination) | - Two-step<br>- Continuous<br>(**C**/50) | - Two-step<br>- Continuous<br>(**C**/10 to **C**/20 with charge termination) |

### Table B-2 Charging Options for Sealed-Lead Batteries
#### (Reference: Section 4.2.5 of text)

**Sealed-Lead Data B**

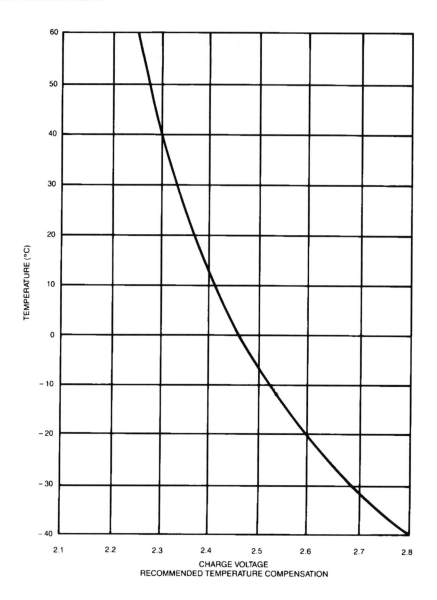

*Figure B-6 Temperature Compensation for Constant-Voltage Charging*
*(Reference: Section 4.2.6.4 of text)*

**Sealed-Lead Data B**

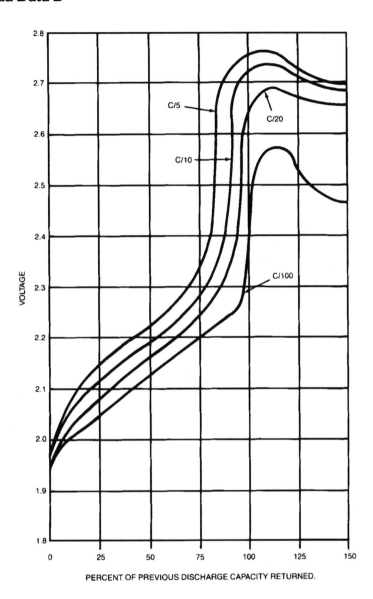

**Figure B-7 Typical Voltage Profiles for Constant-Current Charging**
**(Reference: Section 4.2.7 of text)**

**Sealed-Lead Data B**

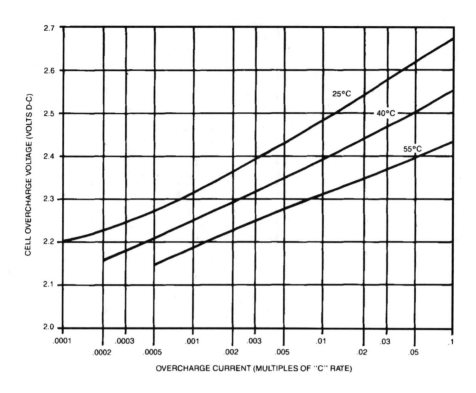

*Figure B-8 Cell Voltage vs. Overcharge Current (Tafel Curve)*
*(Reference: Section 4.2.4.3 of text)*

## Sealed-Lead Data B

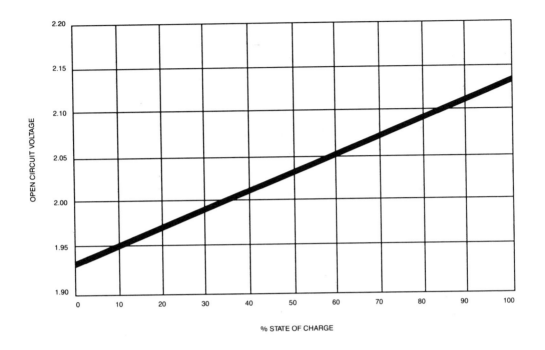

### Figure B-9 Open-Circuit Voltage versus State of Charge
**(Reference: Section 4.3.3 of text)**

# Glossary

The following definitions are not intended to be all-inclusive, but rather to guide the reader in understanding common terms used in relation to Gates Energy Products' batteries.

**A.N.S.I.**   American National Standards Institute.

**ACCUMULATOR**   *See* Secondary Battery.

**ACTIVE MATERIAL**   Specific chemically reactive material at the positive or negative electrode that takes part in the charge and discharge reactions. In the nickel-cadmium cell, nickel hydroxide and cadmium hydroxide are used as active materials at the positive and negative electrodes, respectively. In the sealed-lead cell, lead dioxide and sponge lead are the active materials at the positive and negative electrodes. Because it participates in the cell reaction, the sulfuric-acid electrolyte can also be considered an active material in a sealed-lead cell.

**AGING**   Permanent loss of capacity due either to repeated use or to passage of time.

**ALKALINE BATTERY**   Primary battery which employs alkaline aqueous solution for its electrolyte. Most common usage is in reference to a manganese dioxide-zinc cell with potassium hydroxide electrolyte.

**AMPERE-HOURS**   Product of current in amperes multiplied by the time current is flowing. Capacity of a cell or battery is usually expressed in ampere-hours. Abbreviated as Ah.

**ANODE**   Electrode at which an oxidation reaction (loss of electrons) occurs. In secondary cells, either electrode may become the anode, depending upon direction of current flow. The negative electrode is the anode on discharge.

**ASSEMBLED BATTERY**   Any battery composed of multiple cells.

**AVAILABLE CAPACITY**   *See* Capacity.


**BATTERY**   One or more electrochemical cells electrically interconnected to form one unit and having provisions for external electrical connections.

**BATTERY CASE**   Battery box or enclosure which contains cells, connectors, and associated hardware.

**BOOST CHARGE**   Charging of batteries in storage to maintain their capacity and counter the effects of self-discharge.

**BOUNCING**   Repeated cycling of a battery/charger system in overcharge between fast-charge rate and trickle charge rate. Caused by using an automatic reset thermostat or other switch that does not latch when terminating the fast-charge current.

**BUNSEN VALVE**   One form of resealable vent used on sealed batteries, especially on Gates sealed-lead cells.


**C**   Rated capacity of the cell or battery. Cell charge/discharge current is often specified in terms of a multiple of **C**. (For example the 0.1**C** current for a cell rated at 1.4 Ah is 140mA.) *See also* Capacity, Rated Capacity.

**C RATE**   Charge or discharge current in amperes that is numerically equal to the rated capacity of a cell in ampere-hours.

**CADMIUM ELECTRODE**   *See* Negative Electrode.

**CADMIUM HYDROXIDE**   Active material at the negative electrode of the nickel-cadmium cell.

**CAN**   Metal external covering for cell. For the Gates sealed nickel-cadmium cell, the can serves as the container for the cell as well as being the negative terminal. For the Gates wound sealed-lead cell, the metal can serves only as physical protection of the cell. The cell reactants are contained within the plastic cell jar and the can is electrically neutral.

**CAPACITY**   Amount of charge available from a battery or cell. Measured in ampere-hours. There are various subdefinitions of capacity as listed below depending on the conditions under which it is measured.

> **Actual Capacity**   Cell capacity of a fully formed and fully charged cell when measured under non-standard conditions except using a standard end of discharge voltage.
>
> **Available Capacity**   Capacity of a fully formed and fully charged cell delivered under non-standard conditions including non-standard end of discharge voltage.
>
> **Deliverable Capacity**   *See* Available Capacity.
>
> **Dischargeable Capacity**   Capacity which a partially discharged cell can deliver before it becomes fully discharged or the amount of capacity that can be withdrawn after a limited charge input.
>
> **Nameplate Capacity**   *See* Rated Capacity.
>
> **Nominal Capacity**   Capacity typically delivered under standard conditions; a value greater than rated capacity which is a minimum value.
>
> **Permanent Loss of Capacity**   Reduction in cell capacity from the "fully formed, as new" value, measured under standard rating conditions, that is not recoverable by reconditioning. *See also* Failure.
>
> **Rated Capacity**   Capacity value shown in the specification sheet. The minimum expected capacity when a new, but fully formed, cell is measured under standard conditions. The basis of the C rate. Note that rated capacities depend on the standard conditions used which may vary from battery to battery.
>
> **Recoverable Capacity**   *See* Temporary Loss of Capacity.
>
> **Residual Capacity**   Capacity remaining after discharge.
>
> **Retained Capacity**   Capacity remaining in a cell after an open circuit rest period. The result of self-discharge.
>
> **Standard Capacity**   Cell capacity measured under standard conditions. *See* Standard Conditions.
>
> **Temporary Loss of Capacity**   Reduction in cell capacity that is recovered when the cell is reconditioned. *See also* Failure.
>
> **Useful Capacity**   *See* Available Capacity.

**CAPACITY RECONDITIONING**   *See* Reconditioning.

**CARBON-ZINC CELL**   Primary cell having a zinc anode and a manganese-dioxide cathode with an ammonium-chloride and zinc-chloride aqueous electrolyte.

**CATHODE**   Electrode at which a reduction reaction (gain of electrons) occurs. In secondary cells, either electrode may become the cathode, depending upon direction of current flow. The positive electrode is the cathode on discharge.

**CELL**   Electrochemical unit, composed of positive and negative electrodes, separator, and electrolyte, which is capable of storing electrical energy. When encased in a container and provided with electrical terminals, the cell is the basic "building block" of a battery. Although the capacity of the cell is determined by its size, the cell's voltage is strictly a function of the basic electrochemistry of the couple.

**CELL CASE**   *See* Container.

**CELL JAR**   *See* Container.

**CELL HISTORY**   *See* History.

**CELL POLARITY REVERSAL VOLTAGE**   For a given battery, the minimum end-of-discharge voltage that prevents reversal of any of the cells.

**CELL REVERSAL**   Reversing of polarity of terminals of a cell in a multi-cell battery due to overdischarging.

**CHARGE**   Return of electrical energy to a battery.

**CHARGE ACCEPTANCE**   Willingness of a battery or cell to accept charge. May be affected by cell temperature, charge rate and state of charge.

**CHARGE COLLECTOR**   *See* Current Collector.

**CHARGE EFFICIENCY**   Value obtained when the increase in dischargeable capacity of the battery is divided by the current input. A measure of charge acceptance. Also indicates the amount of the charge input which will not be going into gas generation and cell heating.

**CHARGE RATE**   Rate at which current is input to battery. Various subdefinitions listed below indicate the speed at which the battery is returned to full capacity from a fully discharged state.

> **Fast Charge**   Fastest return of battery to fully charged state. Specifically refers to a class of nickel-cadmium cell specially designed for rapid recharge. With appropriate choice of cell and charger, full recharge may be obtained in significantly less than one hour. Charge rate cannot be maintained in overcharge without damage to cell. Requires both a cell designed to accept charge at high rates and a charger that will deliver high rates until cell is approximately fully charged then switch to a trickle charge rate.

> **Float Charge**   Charging for batteries used in backup applications that reduces the charge rate to prolong life while maintaining the battery in a ready-to-serve condition.

> **Quick Charge**   Highest charge rate that can be maintained indefinitely in overcharge. Specifically refers to a class of specially designed nickel-cadmium cells that are often able to return cells to fully charged state in 3 to 5 hours.

> **Standard Charge**   Charge rate that can be maintained indefinitely without requiring either special cells or switching chargers. Normally returns cell to full charge overnight.

> **Trickle Charge**   Charge rate that will maintain the battery in the fully charged state while reducing overcharge temperature and thereby prolonging life when compared to other charging rates.

**CHARGE RETENTION**   Capacity remaining after a period of storage of a fully charged battery.

**CHARGER**   Device that supplies electrical energy to a secondary battery in a form that can be used to reverse the discharge reactions within the cells.

**CONDITIONING**   Cyclic charging and discharging of a battery to ensure that it is fully formed and fully charged. Sometimes indicated when a battery is first placed in service or returned to service after prolonged storage.

**CONNECTOR**   Device used to make external electrical connections to a battery through mechanical means.

**CONSTANT CURRENT**   Charging method in which current does not change appreciably in magnitude, regardless of battery voltage or temperature. The preferred charging method for sealed nickel-cadmium batteries. Often abbreviated CC.

**CONSTANT POTENTIAL**   Charging method which input voltage does not change appreciably in magnitude regardless of battery state of charge. The most common method of charging for sealed-lead batteries. Often abbreviated CP.

**CONSTANT VOLTAGE**   *See* Constant Potential.

**CONTAINER**   Cell enclosure in which plates, separator, and electrolyte are held. It is normally made up of the cell jar and cover that are permanently joined.

**CONTAMINANT**   Undesirable component, usually chemical, within the cell which reduces its capacity or life.

**CORROSION**   Term often used to describe the gradual oxidation of the metallic lead in the grid for the positive plate into lead dioxide.

**COULOMETER**   Electrochemical or electronic device, capable of integrating current-time, used for charge control and for measurement of charge inputs and discharge outputs. Results usually reported in ampere-hours.

**COUP DE FOUET**   A momentary drop in voltage occurring at the onset of high-rate discharges in some batteries. It is followed by a recovery to the loaded voltage.

**COUPLE**   Combination of anode and cathode materials that engage in electrochemical reactions that will produce current at a voltage defined by the reactions.

**CURRENT COLLECTOR**   Structure within the electrode that allows current to be transmitted between cell terminals and the active materials. *See* Grid.

**CUTOFF VOLTAGE**   Voltage at which a discharge or charge is terminated.

**CYCLE**   Charge followed by a discharge, usually repeated on a regular basis.

**CYCLE LIFE**   Number of cycles a battery survives before its capacity falls below the acceptable level. *See* Failure.

**CYCLING DOWN**   Loss of capacity caused by insufficient charging between repeated discharges.

**CYLINDRICAL CELL**   Battery cells in cylindrical form. Most commonly associated with primary cells and wound sealed nickel-cadmium and sealed-lead cells.

**DEAD BAND**   Range between the temperature at which the thermostat opens and the temperature where it closes (resets). If the temperature first exceeds the point at which the switch opens and then drops below this point, the switch remains open within the "dead band" until the temperature falls below the reset point.

**DEEP CYCLING**   Charge/discharge cycle where approximately 100 per cent of the available capacity is withdrawn at a low rate.

**DEEP DISCHARGE**   Condition where a cell is fully discharged at a low rate resulting in removal of all dischargeable capacity.

**DEPTH OF DISCHARGE**   Capacity removed from a battery divided by its actual capacity, expressed as a percentage.

**DISCHARGE RATE**   Rate at which current is withdrawn from a battery. May be expressed in absolute terms (amps) or in relative terms (as a fraction or multiple of the C rate).

**DISCHARGING**   Withdrawing electrical energy from a battery.

**DRY BATTERY**   Battery that, because of details of its construction, can be shipped without having to meet certain restrictions on transportation of flooded batteries. Both Gates sealed nickel-cadmium and sealed-lead batteries have been classified as dry batteries by appropriate transportation authorities.

**DRY CELL BATTERY**   Term sometimes used to describe a Leclanché or carbon-zinc cell.

**DUMP-TIMED CHARGE (DTC)**   Charging method in which the cell is first discharged to effectively zero capacity and then recharged on a timed charge.

**DUTY CYCLE**   Condition and usage to which a battery is subjected during operation, consisting of charge, overcharge, rest and/or discharge.

**EFFECTIVE INTERNAL RESISTANCE, $R_e$**    Apparent opposition to current flow within a battery that manifests itself as a drop in battery voltage proportional to the discharge current. Its value is dependent upon battery design, state of charge, temperature, and age.

**ELECTRODE**    Conducting body within the cell in which the electrochemical reactions occur. It normally consists of the active material and the structures necessary to collect the charge and to support the active material as required.

**ELECTROLYTE**    Medium, usually liquid, within the cell that permits the movement of ions between electrodes. Nickel-cadmium cells contain an alkaline electrolyte, usually a dilute potassium hydroxide solution. Sealed-lead cells contain a dilute sulfuric acid electrolyte.

**END-OF-CHARGE VOLTAGE (EOCV)**    Voltage of the battery at termination of a charge but before the charge is stopped.

**END-OF-DISCHARGE VOLTAGE (EODV)**    Voltage of the battery at termination of a discharge but while still under load. Standard end-of-discharge voltages that depend on discharge rate have been established for rating purposes.

**ENERGY DENSITY**    Energy stored within a battery or cell as a function of weight (gravimetric energy density — watt-hours per gram) or volume (volumetric energy density — watt-hours per cubic centimeter). Rate dependent.

**ENTRAINMENT**    Process whereby gases generated in the cell carry electrolyte out of the cell.

**ENVIRONMENTAL CONDITIONS**    External circumstances to which a cell or battery may be subjected, such as ambient temperature, humidity, shock, vibration and altitude.

**EQUIVALENT CIRCUIT**    Circuit using conventional lumped parameters that simulates the electrical behavior of a cell.

**EQUIVALENT INTERNAL RESISTANCE**    *See* Effective Internal Resistance.

**EQUIVALENT NO-LOAD VOLTAGE, $E_0$**    Numerical value of the source voltage in the equivalent circuit.


**FADING**    Long-term loss of capability with use.

**FAILURE**    Condition in which a battery is unable to perform satisfactorily. Various forms of failure are described below.

    **Function Failure**    Condition in which the battery causes the end product to fail to perform as expected.

    **Permanent Failure**    Condition where the cell or battery cannot be restored to satisfactory performance.

    **Reversible Failure**    Failure condition which may be corrected through the application of certain electrical procedures or reconditioning.

**FAST-CHARGE**    Charge rate of 1C or greater applied to a battery designed to handle that charge rate. Cannot be used as an overcharge rate so the charging system must switch to trickle charge rate as battery approaches overcharge.

**FAST-CHARGE BATTERY**    Sealed nickel-cadmium battery designed to be charged at the fast-charge rate and which gives a suitable signal which can be used to terminate the fast-charge current without damage to the battery.

**FAST CHARGING**    Rapid return of energy to a battery, usually at the 1C rate or greater.

**FLOAT**    Battery duty cycle (often associated with power backup applications) featuring long periods of time on overcharge and infrequent discharges.

**FLOAT CHARGING**   Charging approach that minimizes the deleterious effects of prolonged overcharge as experienced in float duty. Float charging often consists of constant-potential charging at relatively low voltages for sealed-lead batteries or switched-rate constant-current charging for sealed nickel-cadmium batteries.

**FLOAT LIFE**   Life of a battery measured in calendar time (years) when essentially all of its life is spent in an overcharge condition.

**FLOODED CELL**   Cell where the electrodes are immersed in a pool of electrolyte, thereby eliminating most opportunities for recombination. As a result the cell vents gases through most of its charge cycle. Flooded cell is a term typically used with lead-acid batteries while vented cell is used for the equivalent form of nickel-cadmium battery.

**FLOODING**   Filling the pores of an electrode with electrolyte solution, thereby minimizing the access of gases to the active materials.

**FORM FACTOR**   Geometric shape of battery configurations which may be created by interconnecting cells in various arrangements.

**FORMATION**   Initial electrical charge applied to a sealed-lead cell to convert most of the paste on the plates to active materials.

**FULLY FORMED**   Batteries, especially sealed-lead, that have all of the paste on the plates converted to active materials. *See also* Conditioning.

**FUNCTION FAILURE**   *See* Failure.


**GAS RECOMBINATION**   Method of suppressing hydrogen generation by recombining oxygen gas on the negative electrode as the cell approaches full charge, thereby partially discharging the negative electrode and suppressing hydrogen formation.

**GASSING**   Formation of gas by the plates as the cell approaches full charge. This gas can either be recombined or remain within the cell until the pressure increases to the point where the cell vents. Gassing is a plate-related phenomenon while venting (release of gas to the outside environment from the cell) is a cell-related phenomenon.

**GEMAX™**   Gates high-performance sealed nickel-cadmium cells.

**GOLDTOP®**   Gates high-temperature sealed nickel-cadmium cells that are capable of operating continuously at temperatures up to 70°C.

**GRID**   Framework that supports the active materials within the electrodes. Also serves as the current collector.

**GRID GROWTH**   Increase in dimension of lead battery plates caused by oxidation of the metallic lead grids to lead dioxide which consumes more volume.


**HIGH-RATE CHARGE**   Charge at a rate equal to or greater than 1C.

**HIGH-RATE DISCHARGE**   Discharge at a rate greater than 5C.

**HIGH RESISTANCE SHORT**   *See* Short.

**HISTORY**   Electrical and mechanical environments which a cell has experienced including age, previous use, temperature exposure, charge and discharge.


**I.E.C.**   International Electrochemical Commission

**IMPEDANCE**   AC circuit's apparent opposition to current; consists of reactance and ohmic resistance. For the equivalent phenomenon in a DC battery, see resistance.

**INTERCONNECTIONS**   Method of providing electrical linkage between cells in a battery. May be external in the case of single cells or internal in the case of monobloc batteries.

**INTERMITTENT SHORT**   *See* Short.

**INTERNAL RESISTANCE**   Apparent resistance value calculated from the cell voltage difference between high rate and low rate discharge. *See* Effective Internal Resistance.

**KOH**   Potassium hydroxide.

**LEAD DIOXIDE**   Active material for the positive electrode in a sealed-lead battery.

**LEAD SULFATE**   Chemical compound formed at both positive and negative plates of a lead battery when the battery is discharged.

**LECLANCHÉ CELL**   Carbon-zinc primary cell.

**LIFE**   Duration of satisfactory performance, measured in years (float life) or in the number of charge/discharge cycles (cycle life).

**LOADED STORAGE**   Harmful condition of storing a battery under load, a non-open circuit storage condition.

**LOW-RATE CHARGE**   Charging at a rate that is slightly higher than the self-discharge losses.

**LOW-RATE DISCHARGE**   Less than 0.1C.

**LOW-RESISTANCE SHORT**   *See* Short.

**MAINTENANCE-FREE BATTERY**   Battery that does not require addition of water. Often applied to sealed recombining cells, but also used for flooded batteries carrying excess electrolyte so that water addition is not required over the course of life.

**MAXIMUM-POWER DISCHARGE CURRENT, $I_{mp}$**   Discharge rate at which maximum power is transferred to the external load. Normally this is the discharge rate when the discharge voltage is approximately equal to one-half of $E_o$.

**MEMORY**   Misnomer for voltage depression, referring to apparent loss of capacity on extended overcharge which can be reversed by reconditioning.

**MIDPOINT VOLTAGE**   Battery voltage when 50 per cent of the actual capacity has been delivered.

**MONOBLOC BATTERY**   Sealed-lead battery constructed as an integrated unit rather than assembled from single cells.

**NEGATIVE ELECTRODE**   Electrode which has an electrical potential below that of the other electrode during normal cell operation. The electrode impregnated with cadmium salts is the negative electrode which undergoes chemical oxidation when a nickel-cadmium cell is discharged. The sponge-lead electrode is the negative electrode for the sealed-lead battery.

**NEGATIVE PLATE**   *See* Negative Electrode.

**NET CHARGE ACCEPTANCE**   Sometimes used to describe charging efficiency. Refers to the amount of discharge capacity that can be delivered as the result of a charging input.

**NICKEL ELECTRODE**   *See* Positive Electrode.

**NICKEL HYDROXIDE**   Active material used at the positive electrode of nickel-cadmium cell.

**NOMINAL CAPACITY**   Typical capacity; greater than the rated capacity which is a minimum value.

**NOMINAL VOLTAGE**    Midpoint voltage observed across battery during discharge at a selected rate, usually at the 0.2C or 0.1C rate.

**OPEN-CIRCUIT VOLTAGE**    Voltage of a battery with no load applied to it.

**OPERATING VOLTAGE**    Voltage between the two terminals when a battery is subjected to a load. Usually it is expressed by the voltage of the battery at the 50 per cent discharge point.

**OVERCHARGE**    Normal application of charge current after the battery has reached full charge.

**OVERCHARGE CURRENT**    Charging current flowing to the battery after all the active material has been converted to a charged state.

**OVERCHARGING**    Continuous charge after battery has reached full capacity. In a sealed cell, a result will be increased cell temperature. In a flooded or vented cell, the result will be venting of gases evolved at the electrodes.

**OVERDISCHARGE**    Discharge past the point where the full capacity of the cell has been obtained.

**OUTGASSING**    *See* Gassing.

**OXIDATION**    Release of electrons by the cell's active material to the external circuit. During discharge, cadmium at the negative electrode of the nickel-cadmium cell and sponge lead at the negative electrode of the sealed-lead cell are oxidized.

**OXYGEN EVOLUTION**    Oxygen gas evolves due to electrolysis of water when a battery on charge reaches a certain voltage. This is called the oxygen evolution potential.

**OXYGEN RECOMBINATION**    Electrochemical process in which oxygen generated at the positive plate during overcharge is reacted (reduced) to produce water at the negative plate at the same rate, generating heat.

**PARALLEL**    Electrical term used to describe the interconnection of batteries in which all like terminals are connected together.

**PASTE**    Mixture of various compounds that are applied to positive and negative grids of lead batteries. These pastes are then converted to positive and negative active materials. *See* Formation.

**PERMANENT FAILURE**    *See* Failure.

**PLATEAU**    *See* Voltage Plateau.

**PLATES**    Common term for electrodes.

**POLARITY**    Electrical term used to denote the relative voltage relationship between two electrodes.

**POLARITY REVERSAL**    *See* Cell Reversal.

**POSITIVE ELECTRODE**    Electrode which has an electrical potential higher than that of the other electrode during normal cell operation. The electrode impregnated with nickel salts is the positive electrode which undergoes chemical reduction during discharge of a nickel-cadmium cell while the electrode with lead dioxide as the active material is the positive electrode for the sealed-lead battery.

**POSITIVE PLATE**    *See* Positive Electrode.

**POTASSIUM HYDROXIDE**    Chemical compound which, mixed with pure water in the correct proportions, is the electrolyte solution used in nickel-cadmium cells.

**POWER BACKUP**    Class of battery applications where the battery is used to supply power in case of failure of the AC input from the electrical utility. *See* Float.

**PRIMARY CELL**    Cell designed to be used only once, then discarded. It is not capable of being returned to its original charged state by the application of current. Both carbon-zinc and alkaline cells are primary cells.

**QUICK CHARGE**    Rate of charging which returns full capacity reasonably quickly, but may be sustained into overcharge with a specially designed cell.

**QUICK-CHARGE BATTERY**    Nickel-cadmium battery that can be charged fully in 3 to 5 hours in a simple, constant-current charger and is capable of continuous overcharge at this quick-charge rate.

**RATE**    Amount of current, either charge or discharge current, frequently expressed as a fraction or multiple of the specification rate, **C**.

**RATED CAPACITY**    *See* Capacity.

**RATING**    *See* Capacity.

**RECHARGE**    Return of electrical energy to a battery.

**RECHARGEABLE BATTERY**    *See* Secondary Battery.

**RECOMBINATION**    Chemical reaction of gases at the electrodes to form a non-gaseous product.

**RECONDITIONING**    Charge/discharge cycling regime to eliminate voltage depression or temporary loss of capacity.

**REDUCTION**    Gain of electrons; in a cell, refers to the inward flow of electrons to the active material. During discharge, nickel hydroxide or lead dioxide at the positive plate is reduced to a lower oxidation state.

**REQUIRED APPLICATION VOLTAGE**    Lowest battery voltage that will produce acceptable end-product function.

**RESEALABLE VENT**    In a cell, pertains to a safety vent which is capable of closing after each pressure release, in contrast to the non-resealable "one-shot" vent.

**RESIDUAL CHARGE**    *See* Capacity, Residual.

**RESISTANCE**    *See* Effective Internal Resistance.

**RESIDUAL CHARGE**    *See* Capacity, Residual.

**RESISTANCE**    *See* Effective Internal Resistance.

**REVERSAL**    *See* Cell Reversal.

**REVERSIBLE FAILURE**    *See* Failure.

**REVERSIBLE REACTION**    Chemical change which takes place in either direction, as in the reactions for charging or discharging a secondary battery.

**SAFETY VENT**    Resealable vent which operates to release abnormal gas pressure due to abusive conditions.

**SEALED CELL**    Cell where the internal environment is controlled and isolated from the external atmosphere, often by use of some form of vent valve. Sealed cells often operate (charge) at above-ambient conditions to promote recombination. Since it is sealed, it is free from routine maintenance and can be operated without regard to position. A sealed cell minimizes release of reactants outside the container.

**SECONDARY BATTERY**    Battery system which is capable of repeated use through employing chemical reactions that are reversible; i.e., its discharge capability may be restored by supplying electrical current to recharge the cell.

**SELF-DISCHARGE**    Spontaneous decomposition of battery materials from charged to discharged states.

**SEPARATOR**   Material which provides physical separation and electrical insulation between plates of opposite polarity. In some cells the separator may also be used to absorb excess electrolyte.

**SERIES**   Electrical term used to describe the interconnection of cells or batteries in such a manner that the positive terminal of one cell is connected to the negative terminal of the next cell.

**SHORT**   Condition in a battery where two plates of opposite polarity make electrical contact with each other. Shorts may be of two forms: hard or intermittent.

> **Hard Short**   Short where the current path between the plates is firmly established and the cell is rendered useless.
>
> **Intermittent Short**   Condition where the short path is unstable such as contact between two plates when the cell is moved. An intermittent short can be either low or high resistance. High-resistance shorts may sometimes be burned off by charging at high charge currents; the process is called zapping the cell.

**SPIRAL-WOUND CELL**   Cell formed by taking plates and separator and winding them up in the "jelly roll" configuration typical of both Gates sealed nickel-cadmium and sealed-lead cells.

**SPLIT-RATE CHARGE**   Charging method in which the battery is charged at a high rate and then automatically reduced to a lower charge rate as the battery approaches full charge.

**SPONGE LEAD**   Porous form of metallic lead that serves as the active material at the negative electrode of a lead battery.

**STANDARD CHARGE**   Overnight return of energy to a battery. No special conditions are required of either cell or charger.

**STANDARD CONDITIONS**   Laboratory conditions of rates, times, voltages, and temperatures during charge, rest and discharge.

**STANDBY**   Non-cyclic use of a battery such as in backup power applications (float duty).

**STARVED CELL**   Cell containing little or no free electrolyte solution; as the pore volume of the electrodes and separator are less than fully saturated with electrolyte, evolved gases reach electrode surfaces readily. Relatively high rates of gas recombination are thus achievable.

**STATE OF CHARGE**   Residual capacity expressed as a per cent of fully-charged capacity.

**STORAGE BATTERY**   *See* Secondary Battery.

**SULFATION**   Process occurring in lead batteries that have been stored and allowed to self-discharge for extended periods of time. Large crystals of lead sulfate grow that interfere with function of the active materials.


**TAB**   One form of battery terminal; a flat metal strip protruding from the cell, often containing a hole for wire connection. Routinely used in smaller sealed-lead cells.

**TAFEL CURVE**   Voltage, temperature, and charging rate relationship in overcharge.

**TAPER CHARGER**   Simple, low-cost charger that delivers high charge currents when batteries are discharged then tapers down to lower currents as the battery voltage rises when the battery approaches full charge. Not suitable for sealed nickel-cadmium batteries, this charger is also limited by its sensitivity to input AC line voltage.

**TEMPERATURE**

> **Ambient Temperature**   Average temperature of the battery's surroundings.
>
> **Cell Temperature**   Average temperature of the battery's components.

**TEMPERATURE CUTOFF (TCO)**   Method of switching the charge current flowing to a battery from fast charge to topping charge by a control circuit in the charger that is activated by battery temperature.

**TEMPORARY FAILURE**   *See* Failure.

**TERMINAL**   Location on the cell or battery exterior that is electrically connected to either the positive or negative electrode. May be either a discrete site, such as the positive and negative tabs on sealed-lead cells, or a more generalized location, such as the positive cover and negative can on sealed nickel-cadmium cells.

**THERMAL RUNAWAY**   Condition particularly affecting nickel-cadmium batteries on constant-potential charge at elevated temperature where the battery will attempt to draw increasing amounts of current. Depending on the current limit of the charging system, the result can be heating that jeopardizes the life of the battery.

**TOPPING CHARGE**   Reduced rate charge that completes (tops) the charge on a cell and can be continued in overcharge without damaging the cell.

**TRICKLE CHARGE**   Low-rate charge used to keep battery fully charged after charging at higher rate.

**VENT**   Normally-sealed mechanism which allows the controlled escape of gases from within a cell.

**VENTED CELL**   Heavy-duty flooded nickel-cadmium cell design in which the vent operates at low pressures during the normal duty cycle to release gases generated in overcharge. Vented cells thus require regular water addition to replace that lost through gassing. A vented cell plate pack contains flat plates, separated by a gas barrier and woven nylon separator, completely immersed in electrolyte. Sometimes called a "flooded" cell, although that term is more often applied to similar forms of lead-acid batteries.

**VENTING**   Loss of gas from a sealed cell.

**VOLTAGE CUTOFF (VCO)**   Method of switching the charge current flowing to a battery from fast charge to topping charge by a control circuit in the charger that is activated by battery voltage.

**VOLTAGE DEPRESSION**   Reduction in discharge voltage partially due to effects of long-term overcharge. Often referred to as "memory," this effect can be eliminated by a deep discharge of the cell.

**VOLTAGE LIMIT**   In a charge-controlled battery, limit beyond which battery potential is not permitted to rise.

**VOLTAGE PLATEAU**   Protracted period of very slowly declining voltage that characterizes many discharges from Gates sealed nickel-cadmium and sealed-lead cells. The plateau ordinarily extends from the initial voltage drop at the start of a discharge to the "knee" of the curve where the cell transitions to a rapidly falling voltage at the end of the discharge.

**VOLTAGE-TEMPERATURE CUTOFF (VTCO)**   Method of switching the charge current flowing to a battery from fast charge to topping charge rate by means of a control circuit in the charger that is activated by either battery voltage or temperature.

**WEAR OUT**   Loss of capacity due to normal use.

**WHIPCRACK**   *See* Coup de Fouet.

**WOUND**   Interior cell construction in which plates are coiled into a spiral.

**ZAPPING**   Approach to burning away internal shorts by use of high charge currents.

# Index